Mini Goats

Everything You Need to Know to Keep Miniature Goats in the City, Country, or Suburbs

Sue Weaver

Mini Goats

Project Team
Editor: Amy Deputato
Copy Editor: Joann Woy
Design: Mary Ann Kahn
Index: Elizabeth Walker

i-5 PUBLISHING, LLC™
Chairman: David Fry
Chief Financial Officer: David Katzoff
Chief Digital Officer: Jennifer Black-Glover
Chief Marketing Officer: Beth Freeman Reynolds
Marketing Director: Cameron Triebwasser
General Manager, i-5 Press: Christopher Reggio
Art Director, i-5 Press: Mary Ann Kahn
Senior Editor, i-5 Press: Amy Deputato
Production Director: Laurie Panaggio
Production Manager: Jessica Jaensch

Copyright © 2016 by i-5 Publishing, LLC™

All rights reserved. No part of this book may be reproduced, stored in a retrieval system, or transmitted in any form or by any means, electronic, mechanical, photocopying, recording, or otherwise, without the prior written permission of i-5 Press™, except for the inclusion of brief quotations in an acknowledged review.

Library of Congress Cataloging-in-Publication Data

Names: Weaver, Sue, author.
Title: Mini goats : everything you need to know to keep miniature goats in the city, country, or suburbs / Sue Weaver.

Other titles: Everything you need to know to keep miniature goats in the city, country, or suburbs
Description: Irvine, CA : i-5 Publishing, [2016] | Includes index.
Identifiers: LCCN 2015043464 | ISBN 9781620082072 (softcover)
Subjects: LCSH: Goats. | Miniature livestock.
Classification: LCC SF383 .W35 2016 | DDC 636.3/9--dc23 LC record available at http://lccn.loc.gov/2015043464

This book has been published with the intent to provide accurate and authoritative information in regard to the subject matter within. While every precaution has been taken in the preparation of this book, the author and publisher expressly disclaim any responsibility for any errors, omissions, or adverse effects arising from the use or application of the information contained herein. The techniques and suggestions are used at the reader's discretion and are not to be considered a substitute for veterinary care. If you suspect a medical problem, consult your veterinarian.

i-5 Publishing, LLC™
www.facebook.com/i5press
www.i5publishing.com

Printed and bound in China
19 18 17 16 2 4 6 8 10 9 7 5 3 1

CONTENTS

INTRODUCTION

If you'd like to keep goats, but you don't have a lot of room to spare, think mini goats. Miniature goats, especially Nigerian Dwarf goats, are quickly growing in popularity in North America for numerous good reasons.

- Minis are cute, smart, and personable. They make great pets.
- Minis include Pygmy Goats, Nigerian Dwarfs, Mini Myotonics (Mini Fainters), Miniature Silky Fainting Goats, miniature dairy goats (scaled-down Nubians, Saanens, Alpines, LaManchas, Toggenburgs, and Oberhaslis), Pygoras, and Nigoras. Minis come in shapes, sizes, coat types, and colors galore.
- Their size makes it a snap to transport them, even in a car or a van, so it's easy to take them wherever you like.
- Miniature milkers, like miniature dairy goats and Nigerian Dwarfs, provide about two-thirds as much high-butterfat milk as a big goat and need half the feed.
- It's easy to keep a pair of miniature goats in a fairly small enclosure if you plan it right (we'll show you how). Minis make ideal urban goats.
- Miniature goats are easier to handle than big goats, so they are better choices for children and for adults with limited strength.
- There is a strong market for minis, especially breeds like the Nigerian Dwarf and Miniature Silky Fainting Goat, so it's easy to sell your mini goat's kids to great homes for a decent price. Minis typically give birth to two to five kids.

We came to minis in a roundabout way. We've kept full-size goats since 2012, when a friend who is a breeder of world-class Nubian goats called to say that some uncaring person had abandoned two half-grown mini goats on her farm. The little goats had to be rehomed as quickly as possible. Would we take them? Of course! Thus, Modo, a winsome Mini Myotonic buckling, and Spike, a first-generation Mini Nubian doeling, entered our lives. Not long after that, another friend rescued an aged, ailing Pygmy Goat from an uncertain fate. We took her in and named her Sweetie.

Our original minis were so cute and engaging that two sets of Nigerian Dwarf twins, first Iggy and Ozzy and then Gizmo and Gonzo, soon joined our ranks. Now, we have Eamon, a sweet Nigerian Dwarf buck; Alijah, a Mini Nubian buck; and a Mini Mancha house goat named Dodger. There are more miniature goats in our future. We adore them. Minis are fun!

CHAPTER 1

Goat FAQs

Miniature goats are fun, interesting, and affectionate animals, but before you get one (or more), you should know what you're getting into. Goats are a commitment. Here are some questions that people often ask (we'll elaborate on these points later in this book). Read on!

What can I do with a miniature goat? Depending on the kind of goat you buy, you can breed and milk her, make things from her hair (fiber), or share her by taking her to nursing homes and hospitals to visit the residents and patients. A reasonably big mini goat can carry some of your stuff when you go camping or pull a small wagon. And goats make wonderful pets.

Can I keep just one goat? Goats prefer the companionship of other goats. A goat kept by herself will be sad and lonely and probably call (that's the sound that goats make, also called bleating) a great deal, making noise that your neighbors won't want to hear. It's always best to have at least two goats. If you live in the country or in a city or suburb that allows other animals, most goats are also happy if they have a sheep, pony, horse, alpaca, or other livestock friend for company. Chickens don't provide enough companionship for a goat.

Don't goats stink? No. Goats are naturally clean. The only goats that smell bad are bucks (adult males used for breeding), and then only during rut (breeding season). Does (female goats) and wethers (castrated males) kept in clean quarters are virtually odorless.

Do goats like people? Yes. Unless your goat has never been around people very much or was mistreated before you got her, she'll want to spend her time with you. She'll show her affection by calling when she sees you or wanting you to come hang out with her.

John with some of the minis (left to right): Gonzo, Iggy, and Ozzy (Nigerian Dwarfs); Modo (Mini Myotonic); and Spike (Mini Nubian).

If she really likes you, she'll rub her forehead against you. Because goats become so fond of the people in their lives, don't get a goat and then ignore her or sell her right away; be certain that you want a goat before you get one. It would be perfect if you could keep your goats (or find someone else to take good care of them) for the rest of their lives. Goats are special and deserve good homes.

Are goats good with children? Yes, they are. Goats tend to love human kids. However, you should always supervise toddlers around goats because a goat can easily topple a small child by accident. And goats tend to shove other goats, dogs, and other small creatures out of their way. The occasional goat considers toddlers fair game.

How long does a goat live? Most goats live to be 12 to 14 years old. The world's oldest goat was a Pygmy wether named McGinty who lived in Hampshire, England. McGinty was 22 years and 5 months old when he died.

Is it expensive to keep goats? Not really. Miniature goats are easy to house, and they don't eat a lot of feed. However, in addition to feed and bedding for your goats' house, you'll have to provide dewormers and veterinary care from time to time.

Your biggest cost may be fencing. Goats are inquisitive and intelligent, and they like to get out of their pens and roam around. They're good climbers, so they might hop up and nap on your car or go next door and raid the neighbor's garden. The saying "good fences make good neighbors" is never more appropriate than when keeping goats.

What kind of housing do goats need? Goats aren't picky about where they live. They do need shelter from wind, snow, and rain. It could be a stall in your regular barn or garage, a field shelter (a goat-sized barn with a roof and three sides), or even a really large dog house. They also need room to walk around and exercise in a fenced pen with something to climb on.

Why can't I tie my goat out with a collar and chain? Tethering (tying out) is very dangerous for goats. Your goat could get tangled up and hurt herself. She could knock over her water bucket and get very thirsty. Worst of all, a dog could come along and kill her. It happens to tethered goats all the time.

Can I keep goats in my city or town? That depends on zoning laws where you live. To find out for certain (and do this before

Mini goats can be affectionate, playful friends for human kids.

It's playtime for Eamon (left), a Nigerian Dwarf buckling, and Alijah (right), an F1-generation Mini Nubian buckling.

Most goats live in outdoor shelters, but they can be house-trained for indoor visits. Young Mini LaMancha Dodger enjoys some couch time with Fred the Dachshund.

buying any goats), ask someone at the zoning office in your city's town hall or county courthouse. Many cities and suburbs allow goats as long as you obey applicable zoning regulations, keep their living area clean, and ensure that they don't make too much noise.

Certain breeds of goats are noisier than others, and some individuals of every breed are quieter or noisier than the norm. If you have close neighbors, you'll have to choose your goats very carefully. For example, most Nubians, full-sized or miniature, tend to call loudly because they can be needier than other breeds; although they're cute and endearing, they aren't the best choice for city living. Small, quiet breeds like Pygmy Goats, Nigerian Dwarfs, Mini Myotonics (Mini Fainters), Miniature Silky Fainting Goats, and most of the miniature dairy breeds like Mini Saanens, Mini Alpines, Mini Toggenburgs, Mini Manchas, and Mini Oberhaslis make perfect in-town goats.

Is caring for goats a lot of work? Goats are trusting animals that look to you for their needs. They must be fed and watered twice a day—summer and winter, rain or snow, no exceptions—and you'll need to learn to trim their feet (it's easy to do). They love humans, so your goats will want you to spend some time with them every day. You'll have to clean their stall or shelter when it needs it. If you live in town or in a suburb, your goats' areas will need more frequent cleaning, but don't worry—goats make compact pellets, not big, floppy "goat pies."

If you milk your goat, you're taking on a great responsibility because she must be milked twice a day, every day, throughout lactation. You'll usually need to have her bred once a year and, to be on the safe side, you must be with her when she gives birth.

What do goats eat? They don't eat tin cans like cartoon goats do. Goats won't eat anything that has dirt or manure on it, and they won't drink dirty water. The most important parts of a goat's diet are clean water and good hay. Some goats also need concentrates—which is another word for grains like oats, corn, and barley—or commercial goat feed. They also require a mineral product that they can lick or nibble on whenever they want.

Are they healthy? For the most part, yes. It's important, however, to buy healthy goats from tested herds. Responsible breeders test their stock for two important goat diseases: caprine arthritis encephalitis (CAE) and caseous lymphadenitis (CL), and some also test for Johnes' disease.

What about milking? What do I have to know in advance? You usually have to breed a doe every year if you want to milk her; otherwise she'll produce less and less as her lactation (the period of time, usually 10 months, for which you milk her) progresses until it isn't worth milking her any longer.

Does "come into milk" to feed their kids, not because their owners want milk; this means that you must take your doe to a buck and have her bred, attend the birth of her kids ("kid her out"), and eventually find the kids good homes. Sometimes, you'll find a doe that "milks through," meaning that she keeps giving enough milk that you needn't breed her every year. This used to be the norm, but few present-day breeders select for the ability to milk through.

Playful twins Iggy (left) and Ozzy (right) are still tiny at three weeks old due to being born prematurely.

Sweetie is a rescued Pygmy doe.

Handsome Mini Myotonic wether, Modo, at about five months old.

Dairy does need a 2-month period of down time to rest up before their kids are born, so you won't have fresh milk all year. Carefully handled goat milk, however, freezes nicely.

Sometimes, low-producing does and does in late lactation can be milked just once a day, but high-producing does must be milked twice a day—every day, no exceptions—within an hour or so of the same times every day.

What kind of goat is best? It depends on what you want to do with your goats. If you want milk, choose a dairy breed. If you want a goat that grows mohair or cashmere, you must choose from the fiber-producing breeds like Nigoras and Pygoras. If your children want to show in 4-H, choose a goat that is eligible for the type of classes they want to participate in.

Any goat except for a Myotonic goat can be used for packing or to pull a small wagon. Myotonic goats are "fainters" that stiffen and sometimes fall over when frightened, and they (or you) can get hurt if one of them faints while wearing a pack or harness.

You'll need a doe if you want to milk or to breed more goats. Wethers (castrated males) make wonderful pets, and they're bigger and stronger than females of the same breed, so they work best for packing and driving. Either sex is fine for growing fiber or as pets, but you don't want to keep a buck, especially in town.

Bucks are males that haven't been castrated. They're usually cute, sweet, and friendly, but even miniature bucks are strong and willful. During breeding season, two glands on their foreheads secrete an incredibly stinky substance called musk. At the same time, a buck in rut does weird things, such as making loud gobbling noises, sticking his tongue out and flapping it up and down, and strutting around in a stiff-legged walk. He will even twist his body and spray urine on his face and beard. If you go near a buck in rut, he will "mark" you by rubbing his stinky face against you or even shooting a spray of urine at you. And bucks have good aim!

Should I get a goat with horns? Horns are beautiful, but they're dangerous. Goats with horns get their heads stuck in fences and feeders, and they sometimes use their horns to bully other goats. A horned goat might not mean harm, but by just turning her head at the wrong time she could injure someone. If children want to show in 4-H, they can't exhibit horned goats.

Don't buy a horned goat and think that you can saw her horns off. Horns are part of a goat's skull. They're filled with blood vessels, and if they break or someone cuts them off, the goat will bleed profusely. Removing horns at their base leaves two big holes in the goat's skull. Dehorning a goat, even when done by a veterinarian, is a gory, cruel mess.

Instead, buy a polled goat (one that was born hornless and has small, hair-covered bumps on her head where horns would be) or a goat that was disbudded as a kid. Disbudding is performed by an experienced person using a red-hot disbudding iron to burn and destroy the horn buds on a baby goat's head. It sounds bad, but it's over quickly and it saves the kid (and you) a lot of pain later on.

How much do goats cost? Miniature breeds tend to cost more than full-sized goats, but they're worth it. Expect to pay about $200 to $350 for a registered doeling (a baby female). Depending on where you live, a registered adult doe from a health-tested herd will cost about $200 to $750. Pet goats, especially unregistered goats or wethers, may cost anywhere from $50 to $250.

CHAPTER 2

Choosing a Breed

Before you choose a breed, it's best to assess your needs. Do you want to milk your goats? If so, how much milk will you need? Some breeds give significantly more milk than others. Or are you looking for fiber to spin or sell? Two of our breeds, the Nigora and Pygora, produce mohair, cashmere, and a combination fiber called cashgora. Meat, perhaps? One breed, the Kinder, is a dual-purpose dairy and meat goat. Pets? It's hard to beat Nigerian Dwarfs, Pygmy Goats, Miniature Silkies, and Mini Myotonics for pets, and wethers of the other breeds make charming pets, too.

It's also important to find out what breeds are available where you live. If you have registered goats and you can't or don't want to keep a buck, you'll have to use artificial insemination or find stud service for your does.

You'll have more goats to choose from if you buy a breed that's popular in your area. If you want something rare or different, expect to either travel to get it or pay to have it delivered.

Size may enter into the equation. Some of our breeds are tiny. Others, like Kinders, Pygoras, and Nigoras are small compared to standard breeds but are pretty substantial alongside Nigerian Dwarfs or Pygmies.

Finally, what breed do you really like? Which one makes you sit up and smile? You'll be happier choosing a breed that resonates with you than settling for an alternative out of convenience or some other factor.

Out of Africa

Most of the miniature goats in North America are descended, at least in part, from West African Dwarf Goats (scientists refer to them as WAD goats), the exception being some, but not all, Mini Myotonics. WAD goats have been part of sub-Saharan African life for

Pygoras are personable, attractive mini goats that produce beautiful fiber.

centuries, though their exact origin is uncertain. Photos in books depicting Dr. Albert Schweitzer's mission in Gabon in the early twentieth century depict Nigerian Dwarf-type goats being raised for milk and meat. WAD goats are still an important part of rural life in parts of Africa. There are an estimated 11 million WAD goats in Nigeria alone. They're the most popular type of goat raised in eighteen western and central African countries.

WAD goats vary slightly from country to country, though most are 15½ inches to 22½ inches tall and weigh 44 to 66 pounds. There are two types. One is a cobby, achondroplastic dwarf with a broad body, heavy bone, short legs, and a short, wide head, and most are agouti colored (a "grizzled" appearance with hairs that are banded dark and light): think Pygmy Goats. The other type is a more refined, more normally proportioned animal that is usually black, brown, or butterscotch in color, with or without white spots: think Nigerian Dwarfs. Both are highly fertile, extremely hardy, and resistant to a deadly insect-borne disease called trypanosomosis that kills most other breeds of goat. They are the perfect product of their environment.

The first documented importation of WAD goats occurred in 1909. Additional goats arrived between the 1930s and late 1960s. Consuela Vanderbilt imported WAD goats in the 1940s as eye candy for the Vanderbilts' lavish estates. The National Zoo in Washington, DC, imported sixty goats in the mid-1950s, and, in 1966, William Randolph Hearst imported WAD goats from Cameroon. Zoos snapped up early imports and their offspring, and the University of Oregon Medical School maintained a herd for biomedical research.

Did You Know?

Nigerian Dwarfs and Pygmy Goats have been bred to full-size goats of various breeds to develop more than a dozen distinctive medium-sized to miniature breeds.

At first, all imported WAD goats were called Pygmies. The Animal Research Foundation began registering Pygmy Goats in 1972, and fanciers formed the National Pygmy Goat Association (NPGA) in 1975.

Several early breeders noticed that not all Pygmies were as short-legged and broad-bodied as the rest. The more refined, colorful, dairy-type goats soon came to be known as Nigerian Dwarfs. In 1981, the International Dairy Goat Association (IDGA) established a Nigerian Dwarf breed standard and opened a herd book for these goats. The American Goat Society (AGS) followed with a Nigerian Goat herd book in 1984. In 2002, the American Dairy Goat Association (ADGA), America's premier goat registry, admitted Nigerian Dwarfs to their breed lineup. They sanctioned the first-ever Nigerian Dwarf classes at the 2010 ADGA national show.

Imported WAD goats were originally called "Pygmies," forming the foundation for the modern Pygmy Goat, pictured here.

WHICH REGISTRY?

Some breeds are serviced by more than one registry, and registries don't always honor other organizations' paperwork. Keep this in mind when buying goats.

For example, even though the International Dairy Goat Association (IDGA) opened the first Nigerian Dwarf herd book in 1981, none of the other Nigerian Dwarf registries registers the offspring of Nigerian Dwarfs backed solely with IDGA papers. The Miniature Dairy Goat Association (MDGA) and The Miniature Goat Registry (TMGR) do not accept the offspring of full-size dairy goats registered solely with IDGA.

If you plan to do business with a certain registry, avoid disappointment by carefully studying their registration policies before buying goats; choose animals whose offspring can be registered with that group.

A Nigerian Dwarf.

The Nigerian Dwarf Group

Nigerian Dwarfs played a large role in creating all seven breeds of miniature dairy goat (Mini Alpines, Mini Guernseys, Mini LaManchas, Mini Nubians, Mini Oberhaslis, Mini Saanens, and Mini Toggenburgs) as well as Nigoras and Miniature Silky Fainting Goats.

Nigerian Dwarfs

The Nigerian Dwarf is the fastest growing breed of goat in North America. Nigerian Dwarfs are small, gentle, intelligent, colorful, and productive. Three to five Nigerians fit neatly in the space needed to keep one typical full-sized dairy goat, yet Nigerian does are amazing milkers for their size. At the height of their lactations, does give from 1 to 8 pounds of delicious milk per day. At 2 pounds per quart, that's a pint to a gallon per doe each day. Nigerian milk is richer than milk produced by most other breeds, averaging 6½ percent butterfat.

Folks who breed Nigerians for dairy qualities select for good-sized teats that make hand-milking a breeze. However, not all Nigerian breeders select for dairy qualities, so when buying milkers or potential milkers, make sure that the breeders with whom you're working aren't breeding strictly for pets.

At least four organizations register Nigerian Dwarfs, and not all registries' breed standards call for the same height and weight requirements. Averaging them out, mature Nigerian does are ideally 17 to 19 inches tall, with some registries accepting does up to 21 inches in height. Bucks ideally range from 19 to 21 inches tall, with bucks up to 23 inches acceptable with some groups. Mature weights range from 50 to 75 pounds.

Nigerians have semi-erect ears that stick up and out to the sides, straight to slightly dished profiles, and sturdy bodies with proportionate legs. They have soft, short to medium-length coats in a wide array of colors, including black, brown, white, butterscotch, red, and cream, with or without black or white markings. Most have brown eyes, but blue eyes are fairly common. Some Nigerian Dwarfs have wattles. Most Nigerians are strongly horned, but there are also polled genetics in this breed.

Nigerians breed year round. Does typically give birth to two to four kids. An average kid weighs 2 to 6 pounds at birth and grows quickly. Many Nigerian bucklings are sexually precocious and begin displaying breeding behaviors, like blubbering and stomping, at a week or two of age. To prevent accidental breedings, it's wise to separate most bucklings from does, including their mothers and sisters, by the time they're roughly 10 weeks old.

Miniature Dairy Goats

Miniature dairy goats are primarily registered by two groups: the Miniature Dairy Goat Association (MDGA) and The Miniature Goat Registry (TMGR). First-generation (F1)

WHAT ARE WATTLES?

If you're wondering what those weird pieces of skin dangling from some goats' necks might be, they're wattles, also known as waddles, toggles, or tassels. They serve no known purpose. Wattles are found on both sexes and nearly all dairy breeds, but they're especially common on LaManchas and Swiss breeds such as Alpines, Saanens, and Toggenburgs.

Some people remove their goats' wattles because wattles can interfere with collar placement, and they're not desirable in the show ring. They can be removed in one of two ways: (1) a very snug rubber band placed at the base of each wattle, especially when goats are kids, makes the appendages slough off in a week or two, or (2) some goat owners use disinfected scissors to snip them off. Usually there isn't any bleeding.

The Mini LaMancha is one of the most popular miniature dairy breeds.

offspring are created by breeding a registered Nigerian Dwarf buck to a registered, full-sized dairy-breed doe. After the first generation, F1 does are usually bred to registered miniature dairy-goat bucks of the same breed. Offspring are registered one generation higher than their lowest graded parent, so if an F2 buck is bred to an F5 doe, their offspring will be registered at F3 status. Crossing back to the original breeds is allowed at any time, but the offspring revert to F1 status. In other words, if an F4 Mini LaMancha buck is bred to a registered, full-sized LaMancha doe, they'll produce F1 offspring.

Both the MDGA and TMGR separate goats into three separate herd books. F1 and F2 goats are registered as experimentals; F3, F4, and F5 goats that conform to breed standards become Americans; and F6 and higher goats that meet breed standards are entered in their purebred herd books.

The MDGA and TMGR have different height requirements for the goats that they register, so keep this in mind when buying breeding stock. The MDGA standard calls for a maximum height of 28 inches for does and 29 inches for bucks, period. TMGR's standards are more exacting, providing maximum heights for experimentals, Americans, and purebreds of each breed.

Mini Alpines

Full-sized Alpine goats originated in the French Alps, but they're considered one of several Swiss dairy goat breeds. They came to America in 1922, when Dr. Charles P. DeLangle imported eighteen does and three bucks. Full-sized Alpine does are at least 30 inches tall and weigh around 135 pounds. Males are considerably bigger, in the 34- to 40-inch and 170-pound range. They are hardy, agile, and friendly, and they easily adapt to most climates. Alpines come in a variety of patterns with French names like *cou blanc* ("white neck"—white front quarters and black hindquarters with black

> **What About Full-Sized Does?**
>
> Some full-sized does of the registered dairy breeds are as small as their miniature registered counterparts. They can be a great buy if you find a good one. You can milk her and breed her to a Nigerian Dwarf buck to produce F1-generation miniature dairy goats. Many people do.

A prize-winning second-generation Mini Alpine.

or gray markings on the head) and *chamoisee* (brown with a black face, dorsal stripe, feet, and legs), as well as pied (spotted). They are lean and refined, with erect ears and straight faces. Full-sized Alpine does are excellent milkers; a good one gives 1 to more than 2 gallons of roughly 3½-percent butterfat milk a day, and many Alpine does milk through.

Ideally, Mini Alpines should look like their full-sized counterparts. Mini Alpine does registered with the MDGA can be no more than 28 inches tall; for Mini Alpine bucks, the maximum is 29 inches tall. In TMGR, experimentals can be a maximum of 32 inches for does and 34 inches for bucks; Americans can be 31 inches for does and 33 inches for bucks; and purebreds measure a maximum of 30 inches for does and 32 inches for bucks.

HOW BIG IS A WETHER?

You're probably wondering why wether heights and weights aren't listed in these breed profiles. Because wethers aren't breeding stock, few organizations register them, so specifications for wethers aren't listed in their breed standards.

Wethers usually mature at about the same height as bucks, and they tend to weigh a little more. You can safely assume that your young wether's adult height and weight will conform fairly closely to those of bucks of the same age and breed.

A Mini LaMancha.

Mini LaManchas

One of the most popular miniature dairy breeds is the Mini LaMancha, also known as the Mini Mancha. Full-sized LaManchas are fleshier than goats of the Swiss dairy breeds, and they're famous for their sweet personalities and for producing 1 to 2 gallons of tasty, 4- to 4½-percent butterfat milk per day. They have short, glossy hair and sturdy bodies, and they come in any color or color combination known to goats. Their most unique feature is their very short ears. "Gopher ears" lack cartilage but have a ring of skin around the ear opening. "Elf ears" have erect, triangular ear flaps up to 1 inch long. The LaMancha is the only American Dairy Goat Association-registered dairy breed that was developed in North America.

Mini LaMancha does registered with the MDGA can be no more than 28 inches tall, with a maximum of 29 inches for Mini LaMancha bucks. In TMGR, experimentals can be a maximum of 30 inches for does and 32 inches for bucks; Americans, 29 inches for does and 31 inches for bucks; and purebreds, a maximum of 28 inches for does and 30 inches for bucks.

Hope is a fifth-generation Mini Nubian.

Mini Nubians

Another extremely popular miniature dairy goat breed is the Mini Nubian. Nubian goats originated in England, where they're known as Anglo-Nubians. Early breeders developed them by crossing Jumna Pura, Zaraibi, and Chitral bucks from India and Africa with native British does. Goats imported to California in 1909 and 1913 formed the nucleus of the breed in North America.

Full-sized Nubian goats are elegant, long-eared beauties with Roman noses. Their long, pendulous ears should ideally extend 1 inch beyond the

HOW TALL IS YOUR GOAT?

Measure your goat's height at his withers, using a measuring standard, not a flexible tape. Stand the goat on a firm, flat surface and make sure that his head is held in a relaxed position, not up, down, or to the side. His legs must be set squarely under him, drawn neither too far back nor forward.

You can buy a measuring standard from a goat-supply catalog or somewhere that sells miniature horse equipment. If you don't have one, have a helper hold a yardstick flat across the goat's withers while you measure from the yardstick down to the ground. This works, but it isn't as accurate.

muzzle when held flat along the face. Nubians are heavier bodied than the Swiss breeds, and they have sleek, short-haired coats in any color or color combination.

Full-sized Nubians tend to be large goats: the minimum heights are 30 inches for does and 35 inches for bucks, but many Nubians are much taller. Nubians do better in hot, humid conditions than do the Swiss breeds or even LaManchas, yet they adapt well to cold-winter climates, too.

Some call Nubians the "Jerseys of the goat world." Like Jersey cows, full-sized Nubians give less milk than the other full-sized breeds, but their milk is uncommonly sweet and tasty, with a butterfat content ranging from 4 to 5½ percent.

Goat owners seem to love or hate Nubians, full-sized or miniature, because of their needy personalities and loud, strident voices. If you want goats that love to be with you and will put up a fuss if you aren't around, this is your breed. However, most Nubians are very talkative, so they aren't the best choice for urban living or any situation in which noise would be an issue.

Mini Nubian does registered with the MDGA can be no more than 28 inches tall; for Mini LaMancha bucks, 29 inches is the maximum. In TMGR, does can be no smaller than 21 inches tall, with a maximum height of 29 inches, whereas bucks have a minimum height of 23 inches and a maximum height of 31 inches.

Mini Oberhaslis

The Oberhasli (oh-bur-has-lee) originated in the Berne region of Switzerland and came to the United States in the 1920s. Oberhaslis were originally considered a type of Alpine goat and weren't classified as a separate breed until the 1960s. Full-sized does are between 28 and

Mini Oberhasli Mystic Acres Debbie Jo enjoys a snack with her kids.

32 inches tall; bucks, 30 to 34 inches tall. Oberhaslis are nearly always chamoisee colored with a rich red base coat; a black dorsal stripe, belly, udder, and lower legs; and a nearly black head with white stripes on its sides. Black Oberhaslis are occasionally seen; black does can be registered, but black bucks cannot.

Oberhaslis have erect ears; straight or dished profiles; short, silky hair; and sweet dispositions. At the height of her lactation, a good full-sized Oberhasli doe gives 1 to 2 gallons of roughly 3½-percent butterfat milk a day.

Mini Oberhasli does registered with the MDGA can be no more than 28 inches tall, and Mini Oberhasli bucks cannot be taller than 29 inches. In TMGR, experimentals can be a maximum of 30 inches for does and 32 inches for bucks; Americans, 29 inches for does and 31 inches for bucks; and purebreds, a maximum of 28 inches for does and 30 inches for bucks.

Mini Saanens and Mini Sables

If Nubians are the Jerseys of the goat world, full-size Saanens (SAH-nen or SAW-nen) are the Holsteins. Like Holstein cows, full-sized Saanen does give large quantities of relatively low-butterfat milk (2 to 3 percent) to the tune of 1½ to 3 gallons per day.

Saanens originated in the Saanen Valley of Canton Bern in Switzerland and came to America in 1904. They have erect ears and straight or slightly dished faces, and they're large, heavy-boned goats, with does standing 30 inches or taller and bucks at least 32 inches tall. Like the other Swiss breeds, they don't fare well in hot, humid climates, and their fair skin

A Mini Saanen on the move. Inset: A Mini Sable kid of ¾ Nigerian Dwarf and ¼ full-size Saanen breeding.

ANGORA GOATS

The Angora goat is an ancient Middle Eastern breed dating to the time of Moses. Angoras originated in the Ankara region of Turkey and came to the United States in 1849, when Sultan Abdülmecid I presented seven Angora goats to Dr. James P. Davis. Today, there are three types: modern white Angoras, modern colored Angoras, and traditional Navajo Angoras. All grow long, lustrous locks of fleece up to 12 inches in length and in one of two patterns: ringlets or flat. This fleece is called mohair (angora fiber comes from Angora rabbits). Angora does bred to Nigerian Dwarf or Pygmy bucks produce Nigoras and Pygoras, respectively.

predisposes them to skin cancer. Many Saanen and Sable does can be milked through without being bred every year.

Saanens are always white or cream with pink or olive-colored skin, which is sometimes lightly speckled with black. Saanens of any other color are known as Sables. Full-size Sables are now considered a separate breed and registered in their own herd book. The MDGA registers Mini Sables in the Mini Saanen herd book; TMGR maintains separate herd books for Mini Saanens and Mini Sables.

Mini Saanen and Mini Sable does registered with the MDGA can be no more than 28 inches tall, and Mini Saanen and Mini Sable bucks, 29 inches tall. In TMGR, experimentals can be a maximum of 32 inches for does and 34 inches for bucks; Americans, 31 inches for does and 33 inches for bucks; and purebreds, 30 inches for does and 32 inches for bucks.

Mini Toggenburgs

Toggenburgs (TOG-en-burg), sometimes called "Toggs," are said to be the oldest Swiss breed. They were developed in the Toggenburg Valley of Canton St. Gallen in northeastern Switzerland some 300 years ago. Full-sized Toggenburgs came to America by way of England in 1883 and were subsequently imported in larger numbers than any other Swiss breed.

A second-generation Mini Toggenburg, Cherry Butte Mocha.

Toggenburgs are large goats, with does in the 30- to 32-inch range and bucks standing 34 to 38 inches tall. They have straight or dished faces; alert, erect ears; and medium-length coats. They come in a brown base color ranging from light fawn to dark chocolate, and they always have the same markings: white ears with a dark spot in the middle of each ear; a white stripe on each side of the face, stretching from above each eye to the muzzle; white lower legs; a white triangle on each side of the tail; and a white spot on each side of the throat.

Full-sized Toggenburgs are prolific milkers, and they're noted for their beautiful udders. A typical full-sized doe produces 2 or more gallons of roughly 3½-percent butterfat milk per day, and Toggs often milk through. A Toggenburg doe, GCH Western-Acres Zephyr Rosemary, holds the Guinness World Record for goat-milk production for giving 9,110 pounds of milk, amounting to nearly 1,140 gallons, in a 305-day lactation. She also holds an American Dairy Goat Association title as All-Time Butterfat Record Holder. Toggenburg milk is said to be tangier than milk from other breeds, although this isn't true across the board.

Mini Toggenburg does registered with the MDGA can be no more than 28 inches tall, with bucks 29 inches tall. In TMGR, 28 inches for does and 30 inches for bucks are the preferred heights for experimentals; for Americans, preferred heights are 27 inches for does and 29 inches for bucks; and purebreds should be 26 inches for does and 28 inches for bucks.

Nigoras

If you'd like to milk your goats, but you want them to grow fiber, too, think Nigoras. Breeders create F1-generation Nigoras by crossing registered Nigerian Dwarf bucks with registered full-size Angora does. After the first generation, Nigoras can be bred to other Nigoras, to Angoras, to Nigerian Dwarfs, or even to registered miniature dairy goats of the Swiss breeds (Alpine, Saanen, Sable, Oberhasli, or Toggenburg). Successive generations can't be more than 75 percent dairy breed or Angora to be registered with either of the two Nigora goat registries.

Nigoras are a popular choice for those who want both milk and fiber.

Nigoras come in all colors and are sturdily but elegantly built. They may have erect ears, like Nigerians; ears that droop to the side, like Angoras; or any type in between, but they can't have Nubian-type pendulous ears. Nigora facial profiles are straight to slightly dished. The American Nigora Goat Breeders Association's suggested height range is a minimum of 19 inches and a maximum of 29 inches, measured at the withers. The Nigora Goat Breeders Society doesn't stipulate a standard height.

Nigoras are primarily fiber goats. Individuals produce one of three types of fiber: Type A (Angora type), with mohair falling in long, lustrous, curly or wavy locks up to 6 inches in length; Type B (cashgora type), blending Angora mohair with cashmere; and Type C (cashmere type), consisting of fine, non-lustrous cashmere fiber, 1 to 3 inches in length, overlaid with coarser guard hair. However, because of their Nigerian Dwarf and sometimes miniature dairy goat backgrounds, Nigoras are decent milkers, too, milking 1 to 3 quarts of rich, tasty milk per day.

Miniature Silky Fainting Goats

Miniature Silky Fainting Goats, also known as Mini Silkies, might be world's cutest goats. With their sweeping floor-length coats and eye-concealing bangs, they resemble Silky Terrier dogs, and that's what their originator, Renee Orr, intended when she developed them. To do so, she crossed longhaired Myotonic Fainting Goat bucks with longhaired Nigerian Dwarf does.

Mini Silkies have long coats and sweet expressions.

Mini Silkies come in an array of colors. They have erect ears and dished faces, long bangs, cheek muffs, and smooth, flowing coats that almost touch the ground. Both blue and brown eyes occur in the breed. Most Mini Silkies are horned, although polled goats are fairly common. Maximum heights are 23½ inches for does and 25 inches for bucks. Due to their Myotonic background, some, but not all, Mini Silkies faint, but fainting is not a required trait for registration.

The Pygmy Group

Pygmy Goats were used in developing two breeds of small to medium-sized goat: the Kinder and the Pygora.

Pygmies are the quintessential miniature goat. They are cobby and compact, wide, full-barreled, and muscular, with weight and bulk proportionally greater than that of

Neko is a good example of a Pygmy Goat.

MINIS OVERSEAS

Pygmy Goats are popular throughout Britain and Europe, where "Pygmy" is used to describe both Pygmy and Nigerian Dwarf goats. Britain has an especially active Pygmy Goat club.

Australia is home to several breeds of miniature goat and two breed clubs. The Australian All Breeds Miniature Goat Society (AABMGS) registers Australian Miniature Goats, Elf Goats, Miniature Angoras, Miniature Nubians, Miniature Boers, Miniature Nubians, Miniature Saanens, Nigerian Dwarfs, and Pygmy Goats.

Australian Miniature Goat does measure 60 cm (23½ inches) or less at 3 years of age and older; bucks measure 65½ cm (25¾ inches) at the same age. They're descended from Australian Bush goats. Elf Goat does must be 63½ cm (25 inches) or less at 3 years of age and older; bucks measure 66 cm (26 inches) at the same age. Elf Goats are descended from the same Spanish stock as our LaManchas. Visit the AABMGS's website for detailed histories and standards for each breed.

The Miniature Goat Breeders Association of Australia (MGBA) registers Australian Miniature Goats, Australian Pygmy Goats, and Australian Nigerian Dwarfs. Check the Resources section at the back of this book for website addresses and contact information.

Nigerian Dwarfs and the other miniature dairy breeds. Does of at least 1 year of age and older must be at least 16 inches tall; their maximum height is 22⅜ inches. Bucks of at least 1 year of age must also be at least 16 inches tall, and 23⅝ inches is their upper limit.

Pygmy Goats have erect ears, dished faces, and full coats of straight, medium-length hair. They are hardy, agile, alert, animated, good-natured, and gregarious. Does breed year-round, and multiple births of two to four or even five kids are the norm. Pygmies come in solid black; black with white accents; black, brown, and gray agouti; and black or brown caramel (white to dark tan accented with black or brown).

The Pygmy is the quintessential pet goat, although some owners milk them. Does give from 1 to 2 quarts of milk, ranging from 5 to more than 11 percent butterfat, per day. The National Pygmy Goat Association (NPGA) claims that Pygmy Goat milk is higher in calcium, phosphorus, potassium, and iron than milk from full-sized dairy breeds.

Pygoras

A Pygora is a fiber goat purposely bred to produce fine fiber for hand spinning. Pygoras produce a lofty, soft, fiber that doesn't become coarse as the goat ages. Add in an affectionate, engaging personality; a manageable size; good health; and fleece in a range of colors, and you have the perfect fiber goat.

Katherine Jorgenson developed the Pygora goat in the early 1970s by breeding Pygmy bucks to Angora does to create sturdy goats that she hoped would grow the kind of fiber produced by Navajo Angoras. The Pygora Breeders Association (PBA) was chartered in

Pygoras are alert, curious, and easy to handle.

1987. The only goat that may bear the name "Pygora" is a goat registered with the PBA. In addition, all Pygora goats must conform to the Pygora breed standard, which includes details on proper conformation, colors/patterns, and fleece characteristics.

First-generation Pygoras are created by breeding Pygmy goats registered with the National Pygmy Goat Association to Angora goats registered with the American Angora Goat Breeders Association (AAGBA). To register kids after the F1 generation, both parents must be registered Pygoras or one parent must be a registerd Pygora and the other a registered Angora or Pygmy Goat. They can't, however, be more than 75 percent Pygmy or Angora.

Pygoras are bred specifically for their beautiful fiber.

Pygoras are well-rounded, neither cobby nor angular. They have straight or dished profiles and medium-length, drooping ears, and they come in several colors. They are alert, curious, friendly, and easy to handle, making them a perfect choice for fiber workers who like a gentle, mid-sized goat. Pygora kids weigh about 5 pounds at birth. Adult does average 80 to 120 pounds and must be at least 18 inches tall. Adult bucks and wethers average 75 to 140 pounds and must be at least 23 inches tall.

Achilles, a magnificent Kinder buck.

Like Nigoras, Pygoras grow one of three types of fiber: Type A averages 6 inches in length and hangs in long, lustrous ringlets; it's very fine and feels silky, smooth, and cool to the touch. Type B fiber is cashgora and has characteristics of both mohair and cashmere; it's soft and curly, warm to the touch, and averages 3 to 6 inches length. Type C is a matte undercoat fiber with crimp and a length of 1 to 3 inches. Type C has the finest diameter of the three fleece types and can be as soft as fine cashmere.

Kinders

Another breed developed using Pygmy genetics is the Kinder, a hardy, dual-purpose meat and dairy breed. Its F1 generation is created by breeding registered Pygmy bucks to registered Nubian does. The reverse breeding is acceptable but not recommended. After that, Kinder-to-Kinder breeding is the norm.

Kinder goats have long, wide ears that stick out perpendicular to the sides of the head and extend to the end of the muzzle or farther when held flat against the jawline. Their facial profiles are straight or dished, and they have short, fine-textured coats. Adult does are 20 to 26 inches tall; bucks, a maximum of 28 inches. Most Kinders weigh 100 to 125 pounds. Adult does give 4 to 8 pounds of 5 to 7 percent butterfat milk per day; that equals ½ to 1 full gallon.

Like Pygmy Goats, Kinders are aseasonal breeders, meaning that they breed year-round. Multiple births are the norm, and kids grow rapidly, achieving 70 percent of their adult weight by they time they're a year old. Kinder carcass weight is about 60 percent of live weight, making them excellent meat goats.

Mini Myotonics

Miniature Myotonic Goats, also known as Miniature Fainting Goats, occasionally have Pygmy or Nigerian Dwarf ancestors, but that's the exception, not the norm. Mini Myotonics are the only standalone miniature goat in our lineup of breeds.

Modern Myotonic goats descend at least in part from three does and a buck that lived in Tennessee in the 1880s. Myotonics gradually spread throughout the southern states, where they were known as Tennessee Fainting Goats, Nervous Goats, Stiff-Leg Goats, Scare Goats, and a dozen or so additional colorful names. In 1988, the Livestock Conservancy (then called the American Livestock Breeds Conservancy) added the breed to its Conservation Priority List and officially declared the Myotonic an endangered breed. The breed has recovered nicely since then.

Myotonic goats don't actually faint. They're affected by a genetic disorder called myotonia congenita that, when the goat is startled or scared, causes skeletal muscles, especially in the hindquarters, to contract, hold, and then slowly release. Episodes are painless, and the goat remains awake. In fact, a fainting goat will often continue chewing food that's in her mouth until the stiffness passes.

Myotonic goats are stocky, muscular, and wide in proportion to their height. Adults range in size from 50 to 175 pounds and more. The most common color is black and white, but they come in all colors, patterns, and markings. The average Myotonic is shorthaired, but some have longer, thicker coats. A Myotonic's coat should be straight, not wavy. Myotonics' medium-sized ears are carried horizontally, they have prominent eye sockets, and their faces are usually dished. Most are horned. These are calm, friendly goats, and they're easy keepers. Does breed year-round, and twin and triplet births are the norm.

Three organizations register miniature Myotonics. The Myotonic Goat Registry (MTG) states that minis can be as light as 50 pounds at maturity and as short as 17 inches at the withers. The American Fainting Goat Organization (AFGO) standard calls for does of 3 years of age and older to be a maximum of 22 inches tall, and bucks to be a maximum of 23¾ inches tall. Mini does registered in the International Fainting Goat Association (IFGA) Mini Certification herd book can be no more than 22 inches tall at 3 years of age; for bucks, the maximum height is 23 inches tall.

The Myotonic is the only mini goat breed not classified as having primarily either Nigerian Dwarf or Pygmy lineage.

CHAPTER 3

Get Your Goats

Before you buy goats, make sure that they're right for you. Try to spend time with goats before you commit—this is for the goats' sake as much as your own. Keeping livestock is a huge responsibility.

Are there zoning laws or fencing laws that might prevent you from keeping goats? Will nearby neighbors complain, especially if your goats make noise? Can you provide time, energy, housing, an exercise area, adequate fencing, quality feed, veterinary care, hoof care, and predator protection for your goats? Keep in mind that goats are social creatures that crave the companionship of other goats. You can't keep just one.

When you're sure that you want goats, decide what you want from them. Are registration papers important to you? Will you raise kids or start with adults? How many? How much money are you willing to spend to buy and maintain them? How far will you travel to visit farms and buy your goats?

Be Prepared

Before you bring your goats home, you will need:

- A permit to keep goats if you live where permits are required.
- Safe shelter, bedding, secure fencing, feeders, feed, water containers, and a consistent source of clean drinking water.
- If you plan to breed your goats, you will need safe kidding quarters, material to make mothering pens, and a well-stocked birthing kit.
- If you keep your own buck, he'll need a companion—usually a wether—and strong, secure quarters with stout fences.

REGISTERED OR GRADE?

Are registration papers all that important? It depends. If you're seeking pet wethers, probably not. If you want to show your goats at sanctioned shows, yes, they are. And although it costs more to buy registered goats, their offspring command higher prices, too. Besides, as stockmen say of all livestock species, it costs as much to feed a grade (unregistered) animal as one with papers.

What you may not know is that you can't skip a generation if you want to breed registered goats. That means that you can't buy a doeling with registered parents, not register her, and expect to sell her offspring as registered stock. A few registries have a "native on appearance" herd book for exceptional does (but not bucks) of unknown breeding, but, for the most part, registered goats must have two registered parents of the same breed that are registered with the registry with which you want to do business.

Have your goats' areas securely fenced and gated before you bring them home.

- Halters, leads, and hoof-trimming equipment. If you plan to have fiber goats, you'll need hand or electric shears; you may also need other specialized equipment, such as a milking stand, show halters, or a fitting stand, depending on what you want to do with your goats.
- A first-aid kit and basic medications (we'll talk about that in Chapter 8).
- Most important: phone numbers of several veterinarians who are familiar with goats and who will see patients after hours and on weekends, as well as phone numbers or e-mail addresses of one or two goat mentors who are willing to offer a helping hand as problems arise.

Buying Good Goats

Let's assume you've read Chapter 2, and you've chosen a favorite breed. Let's make it a popular breed like the Pygmy. Where are you going to find them?

Check for "goats for sale" notices. Look on bulletin boards at feed elevators, farm stores, and veterinary practices, or tack up a "Pygmy Goats wanted" notice of your own. Pick up your local paper and read the classified ads. Talk to veterinarians and county extension agents in your buying area. Check out ads in goat magazines like *Dairy Goat Journal*, *United Caprine News*, and *Dwarf and Mini Magazine* (see Resources).

Ozzy, pictured at three weeks old, came from a breeder/raiser of show goats.

Visit a goat show. Chat with exhibitors between classes. State and county fairs host goat shows and so do breed registries. Check the registries' websites for show listings or e-mail or call them for dates and times.

Join goat-related Facebook groups. There are dozens of groups devoted to the miniature goat breeds. It's a great way to find goats for sale, and you'll learn a lot and make new friends.

Find local breeders. Use your favorite search engine to search online for breeders (e.g., "Pygmy goats sale"). Qualify it, if you like, by state (e.g., "Pygmy goats sale Missouri"). If breeders' websites don't offer what you're searching for, call or e-mail them to inquire further. They may know someone who does. Visit registry websites to scope out their online member-breeder directories. Or, phone or e-mail organizations and ask them where to buy their breed of goat where you live.

COUNTY EXTENSION AGENTS ARE A GOAT OWNER'S BEST FRIENDS

The best, most reliable way to learn how to keep goats in the area where you live is to discuss your needs with your county extension agent. What works for goats in Minnesota isn't necessarily right in California, Texas, or Mississippi. He or she is also your best source for reliable information about local feeds, mineral deficiencies, and effective worming procedures—and, an extension agent's services are free. To locate county extension offices in your area, visit the USDA Cooperative Extension System website (http://nifa.usda.gov/extension).

Sale Barn Goats? No Way!

Always buy goats from individuals, not from sale barns. People send sick goats, ornery animals, and problem breeders to sale barns for their salvage value. In most cases, you'll have no contact with individual sellers, who tend to drop off their livestock and leave. The sheer number of sick sheep and goats that pass through a sale barn leads to a buildup of disease organisms. Animals that weren't sick before they arrived at the sale barn are exposed to disease while they're there. If you do succumb to the allure of the sale barn, always quarantine the new goats for at least 30 days.

Finding a Responsible Seller

If you're buying locally, tap into the local goat grapevine. Ask other goat owners who they buy from and who they avoid. Then, after you've narrowed the field to a handful of people selling your type of goat, contact them and arrange to visit their farms. When you arrive, don't be put off if the seller wants to sanitize your shoes. In fact, consider biosecurity precautions a plus.

Look around. Are the goats housed in safe, reasonably clean quarters? Are there droppings in the water tanks, or is the water otherwise dirty? Are the goats eating moldy hay? Are they limping? Do they cough? Do you see runny eyes or noses? Are the goats skinny? Does the seller show you his or her entire herd? Try to see them all, especially any goats related to the ones you came to buy.

Gonzo at two weeks old.

Ask the seller which vaccines and wormers he or she uses. How often does the seller vaccinate and worm his or her goats? Does he or she test for serious diseases, like caprine encephalitis arthritis (CAE), and can you see documentation to prove this?

If the goats appear healthy and well kept, ask to see their registration papers and their health, vaccination, worming, and production records. When buying registered goats, study their papers to make sure that you're getting what you pay for. Registration papers should be issued in the seller's name; otherwise, the

Mini Nubian Alijah at around three weeks old.

seller can't sign a transfer slip so that the papers can be transferred to you.

In some registries, kids must be registered by their breeders. If you buy eligible but as yet unregistered kids, make sure to ask for a fully filled-out and signed registration application and a completed transfer slip that transfers their ownership to you.

Ask the seller about guarantees. Some sellers give them, some don't. If there is any sort of guarantee, get it in writing.

Finally, is the seller willing to work with you after the purchase should questions or problems arise? A responsible seller will be a helpful resource for you long after the sale.

QUARANTINE INCOMING GOATS

Anytime you bring goats home, be they new goats or goats returning from a show or from being bred, plan to quarantine them temporarily away from the rest of your goats. House them in an easy-to-sanitize area at least 50 feet from any other goats or sheep (sheep are prone to the same diseases and parasites as goats) but within sight of the other goats. Worm them, vaccinate them, trim their hooves, and keep them isolated for at least 30 days. Don't forget to also sanitize the conveyance you hauled them home in.

During the quarantine period, feed and care for your other goats first so you can scrub up after handling the quarantined goats. Never go directly from quarantined animals to your other goats. If you can prevent it, don't let dogs, cats, poultry, or other livestock travel between the two groups, either. When the quarantine period is over, sanitize the isolation area and any equipment you've used on the quarantined goats.

HENRY STEWART'S HINTS FOR NEW GOAT OWNERS

These wonderful tips are from a sheep book I discovered while writing *The Backyard Sheep*. It was published in 1898. These tips are as useful today as they were more than 100 years ago, and they are as appropriate for potential goat owners as they are for new shepherds, so I've taken the liberty of substituting goat terms for sheep terms in his text. I've left in references to fleece for those of you considering buying Nigora or Pygora fiber goats. You can't go wrong following this advice.

- Select that breed of goat to which you take a fancy, for what one admires or loves the most, he will give his mind to the most. You won't go wrong on any breed if it is a good animal, well bred and healthy.
- Get no does under three years of age. Young does need better care than a beginner can give.
- Make friends with your goats. "The good goat owner loves his goats, and they will follow him." But they won't follow anyone who ill uses them.
- Don't confine your goats too closely. Don't put them in confinement, but give them an open shed in which they may go as they wish, in or out. They will know to go in when it rains, which is more than some people do.
- Above all, keep their fleeces dry. A wet fleece makes the goat feel cold anytime. A dry one is always warm and comfortable. Above all things, keep your goats clean. Dry litter and plenty of it will keep the floor from smelling.
- Feed regularly, at the same hour every day. Give the goats all the salt they will eat.
- Take good care of the does that are carrying kids. Don't let them get crowded, or chased, or punched by cows, and don't let them get moldy stuff to eat.
- Be patient, kind, watchful, attentive, prompt, thoughtful, and, above all other things, be regular to the hour in feeding and watering. Goats don't, as a rule, carry watches, yet they are watchful and know the time of day... if they are not attended to, they will let you know by their bleating. Don't wait for this, set your times; the goats will soon know them and be particular to be on time every time. A fretful goat will soon be a sick one, and a sick one is apt to be a dead one in a short time. Keep your goats happy, and they will make you happy.

– Adapted from Henry Stewart's *The Domestic Sheep: Its Culture and General Management*, American Sheep Breeder Press, Chicago, 1898.

Evaluating Goats

Spike, a Mini Nubian doe pictured at one year of age, shows good examples of both horns and wattles.

Choose individual goats based on your needs, taking these factors into consideration.

Health: Buy sound, healthy goats. Turn ahead to Chapter 8 to learn how to tell sick goats from healthy ones.

Conformation and type: Ask the organization that registers your breed of choice for a copy of its breed standard. This is a list of points to look for when evaluating that breed of goat.

Sex: If you don't breed goats, don't buy a buck. Most bucks are sweet and charming, but they're unpredictable, and it takes ultra-sturdy housing and fencing to contain them. If you're looking for pets, brush control, or Nigora or Pygora fiber goats, think wethers (castrated male goats). They cost less than breeding stock, and they aren't preoccupied with kids, heat cycles, or rut, making them extra-pleasant goats to have around.

Horns: Decide up front whether you want horned goats. Horns are beautiful, but most goat owners prefer polled or disbudded goats for several reasons. Goats use their horns against one another to establish their places in the herd's social order as well as in everyday quarrels. Injuries are common, and a torn udder is catastrophic. Horned heads don't fit well in most milking stanchions. Horns get caught in fences, where the stuck goat can be injured by herdmates (goats have no sense of fair play) or killed by a predator. And it's no fun for you to be constantly poked by your goats' horns, even though the pokes are accidental. Mixed groups of horned and nonhorned goats tend to coexist nicely in a pasture setting, but they should be housed separately. Horned goats often bully their nonhorned pals when kept in close quarters.

Teeth: Goats have no upper front teeth. They have, instead, an incredibly tough, hard, rigid pad of flesh called a dental palate in the front of the upper jaw.

Pony, an F1 Mini LaMancha, opens wide to show his dental palate.

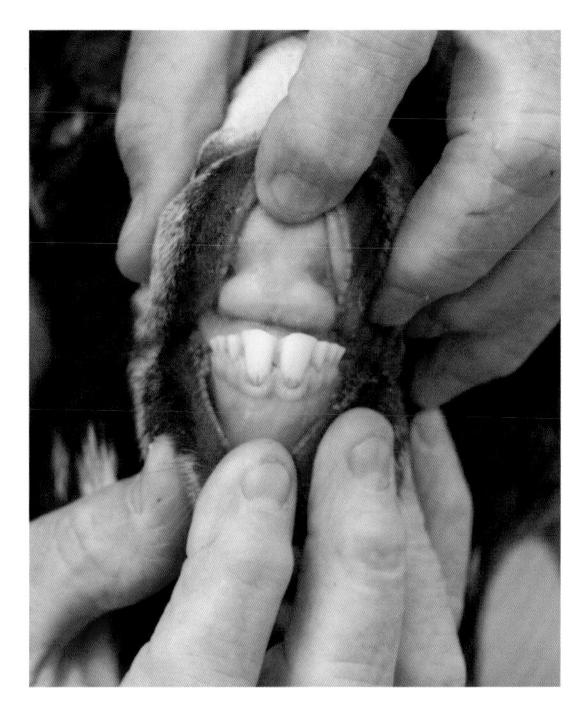
The bottom teeth should meet flush with the dental palate.

A goat's lower teeth should meet flush with her dental pad. If her lower teeth extend beyond the dental pad, she is "sow-mouthed" or "monkey-mouthed;" this is more common in Roman-nosed breeds like the Mini Nubian. If the dental pad extends beyond her lower teeth, she's "parrot-mouthed." Badly sow-mouthed or parrot-mouthed goats have problems grasping and tearing forage when browsing and grazing. They tend to lose weight on pasture and need hay and possibly grain to survive. Both conditions are hereditary and are best avoided.

Age: If you're looking for pets, think about starting with bottle kids. Bottle kids bond with the humans who raise them, so they'll consider you their "mom," and they'll love you. If you're getting into breeding, start with experienced, middle-aged, bred does who can teach you the ropes. Your first kidding season will be scary enough, so don't start with doelings, who are just as confused as you are.

Teat structure: If you're buying a dairy doe or doeling that you plan to milk, you must know what a good udder looks like. We'll talk about this more in Chapter 10. For now, let it suffice to say that goats should have two normal teats—no more and no less. Additional

HOW OLD IS YOUR GOAT?

It's easy to estimate a goat's age by examining the eight teeth in the front of her lower jaw. If all of the teeth are sharp and small, they are baby teeth, and the goat is a kid less than 1 year old. If the center two teeth are big (these are permanent teeth) and the rest are small, the goat is between 1 and 2 years old. If the four center teeth are big and the rest are small, the goat is about 2 years old. If the six center teeth are big and the rest are small, the goat is about 3 years old. If all eight teeth are big, the goat is about 4 years old. After that, the teeth gradually spread and eventually fall out in old age.

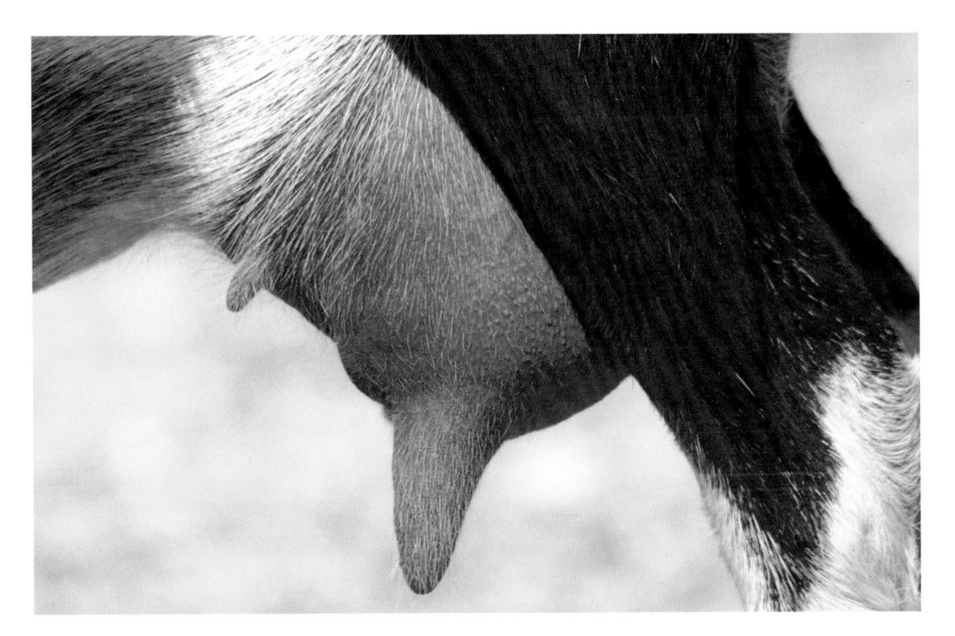

In front of the normal teat is a small, nonfunctional supernumerary teat.

nubbins that don't have an orifice in them (an orifice is the opening that milk flows through) won't interfere with milk production, but teat abnormalities tend to be hereditary, and you won't want your doe to pass this fault along to her kids. Misshapen and double teats are hard, if not impossible, to milk, and kids may not be able to suckle a doe with abnormal teats. When buying a dairy doe or doeling, ask to see her mother's, siblings', and aunts' udders if possible. Also check her father's teats (male goats have teats, too).

Bring Your Goats Home

One of the beauties of miniature goats is that you can haul them in a van, an SUV, or even in the backseat of your car. But unless they're kids and someone is holding them, they should be restrained. It's hard to drive safely when a goat lands in your lap.

The best way to haul miniature and mid-sized goats is in a suitably sized dog crate. Barring that, install the type of barrier designed to keep dogs in the back seat or back area of an SUV. Place a plastic tarp under the crates and over the seats, especially if you haul loose goats. You'll be glad you did.

Goats hauled in airline-style dog crates can also be hauled in the bed of a pickup truck, as can goats in plastic-tarp-wrapped wire crates (but leave the back of the crates uncovered so that the goats can breathe).

You can haul miniature goats in a horse trailer or a "goat tote" (a large metal cage designed to slide into the bed of a pickup truck) but not in a fully enclosed trailer designed for moving furniture or hauling equipment.

SHOPPING FOR YOUR GOATS ONLINE

When you need something goaty, check Craigslist. Craigslist is a series of nationwide online classifieds where people list items for sale. Goats and goat-related goods are usually listed under "Farm and Garden."

Another favorite for buying new and used goat stuff is eBay. Before bidding, be sure to scope out a seller's shipping and return policies, and also check his or her feedback ratings. Be aware that some eBay sellers list goods at eye-catching prices and make up the difference with crazy shipping costs.

When shopping for farm- and livestock-related items, try sites like Jeffers Livestock Supply, Valley Vet Supply, or another of the half-dozen or so semi-discount catalog and online livestock equipment retailers. Be sure to factor in shipping prices when buying online or from catalogs.

When buying online, check out each company's clearance-sale pages. You may find exactly the item you want. Also watch for discount codes and coupons.

Join Freecycle. Freecycle is a great place to find items like cattle lick tubs to use as water troughs, used food-service buckets, old freezers to safely store grain in, and used fencing material. All of the items listed on Freecycle are free.

All of the websites mentioned here are listed in the Resources section at the back of the book.

For times when you need to leave a collar on your goat, buy a lightweight version with a plastic clasp that will break away if the collar gets caught on something.

However you haul your goats, drive carefully, accelerate and brake gradually, and watch those turns. Make sure that there's plenty of slip-proof bedding under the goats' feet. Rubber-backed rugs work well in dog crates and backseats, especially if they're topped with old blankets. Plenty of straw over sand is a good choice for trailers and goat totes.

Make sure that you have your goats' health certificates and health-test results on hand if you buy or show goats out of state. Your veterinarian can tell you what paperwork you need for interstate transport.

GOAT KEEPER'S NOTEBOOK

New Kid in Town

A group of goats can seriously bully a newcomer, so make things easier on the new goat by making sure that she has a friend. For the first few days, place her in a pen or paddock next to your other goats where they can see her, and vice versa, but they can't get together. Next, pick out your friendliest goat and place her in with the new goat until they become friends. Let the herd mingle when the new goat has bonded with her pal.

Familiar Scent

When introducing a new goat to a group of goats, fit her with a collar previously worn by a high-ranking member of the same group. Its scent will help the new goat fit in.

Collars for Goats

If you have to leave collars on goats for any reason, visit the dollar store and stock up on flimsy, single-ply collars with plastic snap closures. The closures are usually strong enough to stay together when leading a goat but will break or give way if the goat snags her collar on a tree limb or stall projection.

Tack Up a Notice

If you need something specific, like a milking stand or another used item, pin notices on bulletin boards in feed stores, veterinary practices, and even places like grocery stores and laundromats. Chances are that someone has just that item but maybe never thought of selling it until reading your notice.

Buy It for Less

If you live near a salvage grocery store, you can save big money on first-aid supplies, vitamins and minerals (grind pills or open capsules to add them to your goat's feed), ingredients for making goat treats, and so much more.

CHAPTER 4

Understanding Goats

Goats can be sweet, wonderful pets or destructive, maddening pests. How you handle and house them makes all the difference in the world. And knowing how to do that means understanding what makes goats tick.

It Makes Sense

It helps to understand how goats perceive their world through sight, hearing, scent, taste, and touch.

Sight

Goats have prominent eyes with panoramic vision in the 320- to 340-degree range and binocular vision of 20 to 40 percent, which means that they have excellent peripheral vision and can see in front and behind themselves without turning their heads. Goats do have blind spots directly in front of and behind them, but, by raising or lowering their heads or moving their heads from side to side, goats can easily scan their entire surroundings. Based on research and on the number of cones and rods in goats' eyes, scientists believe that goats' vision is very keen. Their eyesight is so good that it's hard to sneak up on a goat unless she's dozing.

Goats have horizontally oriented, slit pupils that goat fanciers think are cool and others find rather spooky. When researchers at University of California Berkeley and Durham University in Britain studied the eyes of 214 land species, including goats, they discovered that eye shape is determined by whether an animal is a predator or prey. Predators have circular pupils like our own, whereas horizontally slanted pupils belong to prey. Horizontally slitted pupils provide a much wider field of vision that helps prey animals look out for trouble and stay alive.

All About Goats

Scientific name: *Capra aegagrus hircus*
Average lifespan: 8–12 years
Temperature: 101.5–103.5°F
Heart rate: 60–90 beats per minute; faster for kids
Respiration: 15–25 per minute; faster for kids
Ruminal movements: 1–4 per minute
Length of gestation: 144–155 days
Length of heat cycle: 18–24 days
Length of heat: 12–48 hours
Ovulation: 12–36 hours after onset of standing heat
Number of offspring: One to five (twins and triplets are the norm with most miniature breeds)
Breeding season for seasonal breeders (some Pygoras and Nigoras and most high-percentage miniature dairy goats of the Swiss breeds): August–February
Breeding season for aseasonal breeders (Nigerian Dwarfs, Pygmies, some Myotonics, Kinders, and some miniature dairy goats, especially Mini Nubians): Year-round
Weaning age: 10–14 weeks
Age of sexual maturity (buckling): 10–18 weeks
Age of onset of heat (doeling): 3–12 months

The pupil of a goat's eye is horizontally oriented to give a wide range of vision.

Most goats have brown or amber eyes, but, due to Nigerian Dwarf influence, many miniature goats' eyes are blue. Despite old wives' tales to the contrary, blue eyes are not weaker than brown eyes, and blue-eyed goats can see as well as brown-eyed goats.

Goats aren't color-blind. They perceive colors, but their color vision is not as well developed as it is in humans. A German study conducted in 1980 determined that goats can tell blue, violet, green, yellow, and orange from gray shades of equal brightness.

Hearing

Goats have excellent hearing. Goats with upright ears can direct their ears in the direction of sounds. Goats with long, floppy ears, like Mini Nubians and Kinders, can raise the part of their ears closest to their heads to better hear a specific sound. Goats are frightened by high-pitched and loud noises, like screaming children, barking dogs, or firecrackers.

Scent

Goats have an excellent sense of smell. Smell helps bucks locate does in heat and helps does bond with their kids and distinguish their kids from other does' offspring. Goats also use their sense of smell to locate water and determine differences between feeds and pasture plants.

A goat will *flehmen*, whereby he opens his mouth and curls back his upper lip to transfer scent to a structure called the Jacobson's organ (or vomeronasal organ) located in the roof of his mouth. Bucks frequently flehmen when sniffing the urine of a doe in heat. Both sexes flehmen when they smell something strange.

Alijah, a three-month-old F1 Mini Nubian buckling, with his ears at attention.

The flehmen response helps a goat detect and differentiate between smells.

Taste

Goats have around 15,000 taste buds, whereas humans have about 9,000, so goats taste things more acutely than we do. They prefer certain tastes, particularly sweets and bitters, over others and are surprisingly selective about what they will and will not eat or drink. For instance, unless they're parched, they won't drink water fouled with feces, and they won't eat moldy or musty hay.

Touch

Goats' sense of touch is quite acute. They love it when humans scratch their necks or withers. Flies alighting on a goat elicit a strong response, as anyone who milks goats is sure to agree. Goats respond to touching in many ways, including milk letdown in response to the nuzzling stimulus of their kids or massage by a milker's hands. When young kids sleep, they seek out their mothers and snuggle close beside them.

The long hair along a goat's spine stands up when he's annoyed or, as in this case, when he's being tickled!

THE GOAT SAYS "MEH"

Goats sound off when they are hungry (or think they are), are lonely, or just want to grab your attention. Some does in heat make truly mournful sounds. The average doe moans and grumbles to herself when she's in labor, and she may murmur to her kids even before they're born in a sweet, low, "mama" voice. When heavy labor starts, some does scream; others are more stoic.

Some say that the sound a goat makes is bleating, while others say she's calling. Instead of "baa-ing" like sheep, goats say "meh." Some goats call a lot, while others hardly speak. The amount of noise a goat makes is breed-related, but it varies by individual, too. Nubians and Boer goats are notorious for their loud, strident calls. For the skinny on goat vocalizations, visit www.goats4h.com/GoatSounds.html and listen to goats sound off.

Social Behavior

Something else that will help you understand why your goats behave the way they do is to observe their social hierarchy. Herds of every size from two on up maintain a pecking order or herd hierarchy. One goat leads the pack. All other goats defer to the leader, who is usually a doe that behaviorists call the herd queen. The herd queen sleeps wherever she wants, she gets first dibs on the feed, and when she stares at another goat and says "jump," that goat jumps. The next goat in the social hierarchy defers to the herd queen, but all of the other goats give way to him or her. And so it goes on down the hierarchy to the last goat, the one that all of the other goats pick on whenever they like.

Where a goat stands in the pecking order depends on age, sex, personality, aggressiveness toward other goats, and horn size. Unweaned kids assume their mother's place in the hierarchy and usually rate somewhere just below her after they're weaned.

Low-ranking goats sometimes challenge higher ranking goats and move up the ladder. Newcomers also battle their way into spots in the hierarchy. Fortunately, there is little infighting once a pecking order is established.

Did You Know?

Did you know that goats have accents? Drs. Elodie Briefer and Alan McElligott, researchers at the University of London, say that they do. They found that genetically related kids' cries are very similar but also that kids kept in social groups adjust their calls to match those of their peers. They recorded the calls of week-old kids and then recorded the same kids again when they were 5 weeks old. The older kids had modified their calls to match those of their friends.

When fighting, goats shove and butt each other and side-rake opponents with their horns. Unlike rams, goats don't back up, lower their heads, and race toward their opponents. Instead, battling goats position themselves a few feet apart while facing one another, and then they rear up on their hind legs and swoop down and to the side to smash their heads or horns together.

Goats can bond very closely. Pony (standing) and Tinker are F1 Mini LaManchas who were raised together as bottle babies.

Other forms of aggression include horn threats (chin down with forehead or horns jutting forward), staring, pressing the horns or forehead against another goat, rearing without swooping and butting, and ramming another goat's rear end or side. An angry goat may stand his hair on end, particularly the longer hair along his spine. Goats also express dominance by pawing with a stiff front leg, flapping the tongue, and blubbering. Goats sometimes pull these stunts with humans, too. Don't let them. You want to be top goat in your herd's pecking order.

Goats become more aggressive toward one another when confined in close spaces. Don't confine timid goats with especially aggressive ones or horned goats with hornless ones. Someone will get hurt.

KEEP AN EYE ON THE KIDS

Your human kids, that is. Small children are naturally attracted to barnyard activity, but they need to be supervised when playing around goats. A frightened goat can easily bowl over a toddler, and a doe with newborns who wouldn't think of butting you might consider your 6-year-old fair game. Also, goats high in their herd's social hierarchy may consider small children part of their herd and "put them in their place."

The best thing to do when your children interact with your goats is secure any individuals who habitually challenge children who are too small to correct them. Older children can carry a squirt bottle and shoot water in the faces of goats who try to harass them. If you have a buck, his pen should be off-limits to anyone not old enough or strong enough to protect him- or herself from aggression or amorous advances.

These full-size goats are demonstrating the typical fighting behavior of rearing before swooping down on the opponent.

High-ranking, bossy goats sometimes bully shy, low-ranking goats, even in larger spaces. There's little a goat owner can do to stop this short of separating those two goats, but it's a short-term fix because bullying resumes once the goats are reunited.

Once your goats have formed their pecking order, it's best to maintain the status quo to minimize infighting. Goats tend to maintain family groups and establish close friendships, so try to keep those goats together. Constant regrouping causes stress that can lead to lower milk production and stress-related illnesses.

Play Behavior

Goats are a lying-out species, like deer and cattle, which means that after kidding, feral does stash their newborns in a safe place and then come back periodically to feed them. Some domestic does do this, but others don't. Goat kids are more independent than, say, lambs; lambs shadow their mothers from birth. A kid may seem to be missing when he's actually relaxing in a hidey hole, watching your frantic search. Even does sometimes lose their kids. Be prepared for this and don't panic.

As kids grow older, they venture farther from their mothers and form gangs. Play behaviors include mounting one another (doelings do this, too), play-fighting, leaping on and off of rocks or dirt piles, and springing into the air while ninja-kicking their heels.

Goats love to play. They especially love to climb. All goats descend from the bezoar ibex, a type of mountain goat domesticated in the Middle East about 10,000 years ago and again in a separate domestication event in Pakistan about 9,000 years ago, so goats are genetically wired to climb. Climbing apparatus gives them a chance to play while providing great exercise. Cable spools, downed tree trunks, picnic tables, piles of large

LIVING WITH HORNS

It isn't the end of the world if you've chosen a goat with horns. It just means that you have to learn to work around them. Be aware of your goat's horns when you're working near her and be prepared to suffer some accidental bruises. It helps to cut a small hole in two tennis balls and stick one on each of her horns when you're trimming hooves, giving shots, and the like. You can also temporarily duct-tape tennis balls to the ends of her horns if she bullies a newcomer but only until they work out their places in the new pecking order.

Some horned goats, especially young ones, repeatedly stick their heads through openings in fences and stall dividers and then can't get them out. Your goat may outgrow this habit, but don't count on it. If you have such a goat, fit her with the type of horn guard shown in this picture. Use PVC pipe and exceptionally strong duct tape, such as Gorilla Tape. Wrap it around many times, or else she'll get it off.

Don't haul your goat around by her horns. Goats hate that, and the horns can break off. If you must hold her by her horns, keep your hands down as close to her skull as you can. And in case a horn breaks off, add a strong blood-stop product to your first-aid kit. The kind that comes in a shaker bottle and is sold at the feed store isn't sufficient; get the kind designed for staunching human wounds. Broken horns are a gory mess. If you can't stop the bleeding, call your vet.

rocks, homemade balance beams, and stacked straw bales are fun for goats to climb on. Or provide a ramp so that mini goats can climb to the rooftops of their shelters. Playground equipment and toys of every kind contribute to goats' welfare, especially for urban goats kept in smallish quarters.

Some goats play with sturdy horse and goat toys designed to deliver treats as they're pushed and flung around. You can make your own using large, clean, uncapped plastic jugs with or without treats inside. Tie them to walls or fences or drop them on the ground. Or hang hard rubber dog toys from walls, ceilings, or fences. Children's plastic play houses, teeter totters, and slides are fun to climb on and hide behind or under. Empty cardboard boxes and paper feed bags with staples and plastic liners removed make wonderful temporary toys as long

Like their distant ancestors, goats like to climb on everything.

CABLE SPOOLS FOR CHEAP, EASY FUN

One of the cheapest but most rewarding pieces of playground equipment you can give your goats are cable spools—the big, wooden spools that heavy electrical cable comes on. Goats enjoy climbing and snoozing atop cable spools placed in their outdoor living quarters, so get several. If you ask your local electric co-op for spools, they'll probably give you as many as you can haul away.

Cover all holes with pieces of scrap wood before allowing your goats to play on them. Otherwise, a goat's leg can slip down into a hole, and, as the goat pitches forward, she could break her leg.

Some spools are too big for smaller miniature goats to easily hop up on. If that's the case, make a simple ramp on one side by securing one end of a sturdy plank to the top of the spool. You can also connect two or more spools with balance-beam planks.

as your goats don't consume them—some goats eat paper and cardboard, and although a little paper won't hurt a goat, you don't want her to eat a lot of it.

Feeding Behavior

Goats are browsers, not grazers, which means that they prefer to range over a large area while sampling select weeds, shoots, brush, twigs, and bark rather than mowing down grass like horses or cattle do; however, goats do graze when they can't browse. Goats prefer to nibble the tops of plants instead of devouring them down to the ground, and they'll refuse anything fouled with urine or droppings.

Feral goats stay on the move, feeding, for up to 12 hours a day. Domestic goats kept in range conditions do this, too. They tend to travel in family groups or groups of friends within the overall herd. When they aren't moving and feeding, they nap or ruminate (chew their cud).

Goats prefer to sleep and ruminate on high ground, moving downhill to browse during the day. They generally feed in the early morning and again late in the afternoon, resting in their feeding area between grazing sessions. During periods of extreme heat and humidity, they often rest from mid-morning to evening and feed in late evening and at night. Time spent feeding is affected by weather, quality and availability of browse and supplemental feed, and length of day.

Goats usually ruminate while resting on their sternums (lying down with one or both front feet tucked under their bodies) but also sometimes standing up. The time spent ruminating is about equal to or slightly less than the time spent browsing or grazing. Rumination induces a state of blissful drowsiness.

When goats are kept in small areas, they tend to strip the leaves and bark from trees. If you can't fence your goats away from valued trees, wrap the trees' trunks with several layers of chicken wire as far down and as far up as your goats browse. Or bring your goats brush of your choosing so they eat that instead of your trees. Just be certain that you aren't bringing them species that are poisonous to goats (see Chapter 5).

Miscellaneous Behaviors

We'll talk about breeding and kidding behavior in Chapter 9. Following are some other behaviors to be aware of:

- Kids kneel to nurse, and they bunt their mothers' udders to facilitate milk letdown. A rapidly wagging tail means that a kid is suckling milk. After feeding, contented kids take naps. A kid that constantly calls, suckles, or probes at his mother's udder isn't finding enough to eat.
- Both kids and adult goats examine their world by "tasting" things. Expect them to chew. Watch out for your hair!
- Does allow their own kids to jump and climb on them but usually do not let other goats' kids do the same.

Ozzy helps himself to some hay from the feed cart.

SMART GOATS

How smart are goats? Very. You'll see how true this is when you work with your goats every day. But it's not just us goat owners who think our goats are intelligent. Science agrees.

Researchers at Queen Mary University in London tested goats' IQs by training a dozen goats at Britain's Buttercups Sanctuary for Goats to use a relatively difficult, two-step process to retrieve a pasta and grass reward. To do so, they had to use their lips to pull a string that pulled out a lever, and then they used their muzzles to nudge the lever upward to release the treat.

Nine of the goats quickly figured out the process, averaging twelve tries before they earned the tasty treat—considerably faster than chimpanzees tested with the same equipment. Two goats were disqualified because they used their horns instead of their lips to raise the lever. One goat tried hard but gave up after 22 tries.

The researchers repeated the same experiment 10 months later. All nine goats retrieved their treats on their very first try, averaging 38 seconds from start to finish. One goat earned her treat in 6 seconds flat, and another was so eager to try that she leapt over a wall to access the testing box. Said researcher Dr. Allen McElligott, "[Goats] have the ability to learn complex tasks and remember them for a long time."

- Kids play by head-butting one another. Pressing their foreheads against other goats or people shows aggression, too. This sets up a precedent you won't be happy with when cute little kids grow into adults, so don't let them do it to you. And never push on a goat's head, even in play. If you don't want goats to butt you, don't make it a game while they're small.
- When you have to handle kids (and some adults) for unpleasant procedures, such as vaccinations, they shriek a high-pitched distress call that's sure to bring close neighbors out in droves. Some kids, especially Mini Nubians, sound like human children being tortured. You'll be shocked the first time you hear this bloodcurdling scream.
- When a goat is startled, she will stamp one front foot and sound an alarm that sounds like a human sneeze. Speaking of sneezing, goats are sometimes startled when humans sneeze, so be forewarned.
- Goats are notoriously hard to drive, but they're very, very good about following a leader. If they accept you as the leader of their pecking order, get out front and let them follow you. If they still won't follow, a pail of grain or other goodies makes you herd queen for a day.
- Calm goats move forward at a walk; frightened goats tend to scatter. Keep things low-key to save yourself exasperation and wasted time.

Handling Goats

Handling goats is easier if you know why they do the things they do. For example, all goats have personal security, or flight, zones. Anything scary invading a goat's personal space will cause her to run away. A goat's flight zone might be 50 feet or nothing at all; breed, gender, tameness, training, and degree of perceived threat enter each equation.

A goat's point of balance is at her shoulder, at a 90-degree angle from her spine. If a goat detects movement within her flight zone and behind her point of balance, she will move forward; movement within her flight zone and in front of her point of balance makes her turn and move away. Learn to position yourself to move your goats in the direction you want them to go.

It's much, much easier to move goats by leading them with a bucket of feed than by trying to drive them. In fact, it's next to impossible to drive goats, even with a herding dog, because they don't have the flocking instinct that causes frightened sheep to mob together and move as a group. Instead, a startled group of goats scatters, although some individuals may whirl to face whatever spooked them, sneeze-alarm, and stamp their feet. If they're still alarmed, they run, calling upon their speed and agility to outmaneuver perceived danger.

Other good things to know:

- Goats don't like to cross water or move through narrow openings.
- Goats move uphill more readily than downhill and prefer to move into the wind instead of against it.
- When approaching goats, don't look directly into their eyes; predators do that, and it makes goats nervous.
- Goats that aren't used to being handled can be hard to catch. To catch one, use feed to lure the goat and her friends into a fairly confined space. Stick your arms straight out at your sides to create a visual barrier. Approach slowly and calmly. Quietly ease the goat you want into a corner. If she isn't wearing a collar, quickly place one hand under her jaw and lift up on her head while cradling her rump below her tail with your other hand. Get a collar and lead on her as quickly as you can.
- Any time a goat hesitates and won't move forward, lightly grasp her tail and gently lift it up and forward.

This is an easy way to restrain a mini goat when you need to.

Mini goats have endearing personalities that bring joy to their keepers.

- Don't try to push a goat, especially to the side. Goats resist being pushed by leaning into pressure. Instead, grasp the goat's collar and pull her toward you.
- Goats occasionally slip into a catatonic state when they're deeply frightened. The right thing to do is back off and allow the goat time to recover. When she has, proceed with whatever you were doing in a quieter, more goat-friendly manner.
- Reward goats by scratching them. Scratch a goat's neck, along her spine and sides, her chest, and her rump. Most goats don't care to have their faces petted.
- Goats do almost anything for food, so explore treat-based training methods like clicker training (see Chapter 10) rather than forcing a goat to do something she'd rather not do. Clicker training protocols for dogs and horses are easily adapted for training goats.
- Goats hate to get wet. Therefore, a great way to discipline goats for jumping up on you, showing aggression, or climbing on the roof of your car is to spray them with a strong stream of water from a spray bottle or a squirt gun. Say "no!" at the same time. They'll get the picture.
- When a buckling is born, his penis is secured inside his sheath (the external covering that protects his penis) by a fold of tissue called a frenulum. Bucklings begin trying to break their penises free within a week or so of birth. They do this by curling their hips forward under them and making hip thrusts. It looks strange, but it's normal behavior.

GOAT KEEPER'S NOTEBOOK

Easy Crowd Control
Fasten short leads to your fence or a wall in your goat shed and clip them to your goats' collars when feeding them or cleaning their stalls. Attach leads with breakaway snaps or use flimsy collars that will break if a goat gets into danger while she's tied.

Who's Been Treated and Who Hasn't
When you have to worm, vaccinate, or trim the hooves of several goats that are colored alike, buy an equal number of dollar-store dog collars and, as you finish with each goat, snap a collar around her neck.

Take a Stand
Teach your goats to hop up on a mini-size milking stand or fitting stand where you can secure their heads. It's so handy for doctoring, worming, tattooing, and hoof trimming and can really save your back (and your temper).

Be There
For safety's sake, never use a rope with a slip knot in it to tie up your goat. And don't go off and leave a semi-trained goat tied up; you should be there to save her if she panics.

Don't Get Hurt
Never underestimate the power of a goat, even a miniature goat. When handling frightened goats or working with goats in close quarters, wear long-sleeved shirts, long pants, and boots or steel-toed shoes. Keep small children out of the action altogether.

Water Works
Reward-based training works best, but when you need to zap undesirable behavior (jumping up against you and tap-dancing on your car come to mind) right away, reach for a high-powered water gun or a household pump sprayer with a long, strong jet. Goats hate getting wet, especially when something is squirted in their faces. A loud "no!" coupled with a blast or two of water grabs even the most stubborn goat's attention.

No Chasing
Don't deal with undesirable behavior by yelling and waving your arms and chasing your goat away. To most goats, chasing is play behavior. You're rewarding your goat for misbehaving when you chase her.

CHAPTER 5

Feeding Goats

Goats don't process food the way we do. They, along with sheep, cattle, deer, moose, elk, buffalo, bison, giraffes, and camels, to name a few, are ruminants. They have four-compartment stomachs, and they ruminate, which means that they chew their cud.

The first compartment of the stomach is called the rumen. The rumen is essentially a large fermentation vat where billions of microbes, including molds, yeasts, fungi, protozoa, and bacteria, feed on carbohydrates in the food that a goat eats and convert them to volatile fatty acids. Fatty acids are the goat's main source of energy.

One type of rumen microbe digests fiber. Fiber means forage—browse, grass, or hay—which does best in a 7.0 pH environment. Another type of microbe digests sugars and starches (grain and commercial bagged feed), which prefer a more acidic 5.0 to 6.0 pH environment.

Feed mixed with saliva enters the rumen and begins separating into layers of solid and liquid material. Later, when the goat is resting, he burps up a bolus of food (cud) and gas and then re-chews the solids. When he's finished, he swallows the glob again.

What a goat eats affects the pH of his rumen. When he eats forage, he cuds more often and makes more saliva. This is how it should be. Because grain is much less fibrous than forage, he doesn't have to chew it as much, so he produces less saliva. When a goat eats a lot of grain, his forage-digesting microbes die off, clearing the way for lactic acid-producing microbes to take over. This can cause a serious, sometimes fatal, metabolic disease called acidosis.

Chewed-up feed slurry flows back and forth between the rumen and the next compartment of the goat's stomach, the reticulum, through an overflow flap. It sloshes back and forth for 20 to 48 hours before moving through a short passage into the third compartment, the omasum.

ACIDOSIS

Acidosis, also called lactic acidosis or grain overload, occurs as a goat's rumen pH falls below about 5.5, when rumen microbes begin dying and ruminal action slows down or stops altogether. Symptoms include depression, dehydration, bloat, racing pulse and respiration, staggering, and coma, which lead to death. Due to permanent damage to the lining of the rumen and intestines, a survivor generally fails to thrive. If you suspect that your goat has acidosis, call your vet without delay.

To counter acidosis, does and doelings eating grain should be provided with free-choice baking soda to nibble as the need arises. Don't feed baking soda to wethers and bucks because it is said to contribute to the formation of urinary calculi.

The omasum, sometimes called the manyplies, is divided by long folds of tissue and lined with tiny finger-like projections to increase its working surface. Food particles become smaller as the omasum removes excess fluid from the rumen slurry. The omasum also absorbs volatile fatty acids that weren't absorbed in the rumen.

The final compartment, the abomasum, is the most like a human's stomach. Its walls secrete digestive enzymes and hydrochloric acid. Proteins are partially digested in the abomasum before the semi-digested material flows into the small intestine. There, it's mixed with secretions from the liver and pancreas, which enable intestinal enzymes to break down the remaining proteins into amino acids, complex fats into fatty acids, and starch into glucose. Contractions continually push small amounts of material into the large intestine, where bacteria wrap up digestion, and then out through the anus.

Sweetie, pictured here in her later years, loved watermelon from the garden.

Why is this important to know? Because it shows that the food a goat ingests isn't what truly feeds her. Instead, the things she eats feed her rumen microbes, and those microbes begin the process that ultimately nourishes the goat. That's why it's essential to keep your goats' rumen microbes functioning properly, and the best way to do that is to feed them a goat's natural diet: forage.

Feeding Basics

No matter what you feed your goats, certain rules apply across the board.

- Buy the best feed that you can afford.
- Don't switch feeds overnight. Add a little new feed to a little less of what

your goats are accustomed to, changing gradually over at least a week's time. This allows your goats' rumen microbes time to adjust.

- Feed at the same times every day. Goats fret when meals are late.
- Feed hay in feeders instead of off the ground. Feeding off the ground results in lots of waste and contributes to worm infestation. Choose sturdy, safe feeders that your goats can't demolish or hurt themselves on.

Never give your goats:

- Moldy, musty, dusty, or nasty-smelling feed or feed with insects or dead animals in it. If there's even a hint of animal tissue—a desiccated snake, a decayed mouse—throw the entire bale away.
- Cattle feed containing urea, including some mineral licks marketed for cattle.
- Dog or cat kibble or any other high-protein, meat-based animal feeds.
- Cut-and-carry flowers, leaves, brush, or any other plants that you can't identify and could be poisonous to goats.
- Chemical-treated plants or grass clippings.

Forage

Goats are designed to thrive on cellulose found in forage: weeds, leaves, twigs, grass, and hay. In fact, if your goats are adult wethers or nonproducing does, they need nothing but high-quality forage and a goat-specific mineral formulated for your part of the country. The key words are *high-quality forage*. Here's what you need to know.

Goats need the cellulose found in high-quality forage for healthy digestion.

GOAT TREATS

Goats love treats. The trick to treating your goats is handing out yummies in moderation. Slice or break up large items into smallish pieces because some goats are gobblers and can easily choke. Here are some finger foods that goats enjoy:

- Raisins and dried cranberries
- Peanuts (shelled or unsalted in the shell) or any other type of shelled nuts
- Popped popcorn
- Tortilla chips
- Diced fresh apples
- Diced dehydrated apples, pineapple, or apricots
- Watermelon or cantaloupe rind cut into bite-sized pieces
- Plain or frozen grapes
- Sliced carrots

The following goodies are not as healthy and should be fed in moderation or saved for special occasions:

- Dry breakfast cereal of all kinds
- Animal crackers
- Gummy bears, jelly beans, and similar bite-sized candy (but don't feed chocolate to goats!)
- Pinwheel mints broken into several pieces
- Miniature marshmallows

For something extra special, look for horse treats. If horses like it, goats will love it, too.

Pasture

Goats love to roam woody pasture, selecting the plants they most like to eat. Given the chance, goats consume about 60 percent browse and 40 percent grass. The downside to pasturing your goats is that you need very good fences to keep goats in and goat-killing predators out. The amount of land and the cost of fencing materials and labor may limit the amount of pasture available to your goats.

Hay

Quality hay is an excellent substitute for pasture. There are two kinds of hay: legume hay (alfalfa, red clover, white clover, vetch, birdsfoot trefoil, peanut, soybean, lespedeza, cow pea) and grass hay (Bermuda grass, brome grass, timothy, orchard grass, bahia grass, ryegrass, fescue, bluestem, Reed canary grass, Kentucky bluegrass, native grasses). Native grasses is a nice way of saying "whatever grasses and weeds grow wild in your locale." Hay is often a mixture of two or more varieties of grass and legumes, such as an alfalfa/

timothy mix or orchard grass and clover. However, not all varieties of hay plants grow everywhere in North America. Your cooperative extension agent can tell you what types haymakers put up in your locale and which ones are best for goats.

Choosing Hay

Miniature goats don't eat a lot of anything, so it won't break the bank to buy the best feed you can find. How much hay you'll need depends on the size of your goats, what you do with them (milking, pets, breeding stock), the quality and variety of hay you choose, and how long your hay-feeding season might be.

Hay is cheapest when purchased in big, round bales weighing 600 to 1,500 pounds. However, unless you have a lot of goats, large bales are probably not your best option. To preserve quality, a big bale should be consumed in a week or less when fed outdoors and exposed to the elements. The only time feeding big bales is a wise choice is when you have a place to store it indoors, away from your goats, and you can fork off hay and feed a little as needed.

Your best bet is to feed "small square bales," although they're actually rectangular rather than square. These bales weigh anywhere from 40 to 120 pounds, they're tied with two strands of twine or wire, and they easily separate into easy-to-feed sections called flakes. Many feed and farm stores sell small square bales of hay, so you can buy them as you need them, which is a boon when feeding a few small goats.

In some places, you can also buy hay in big square bales that range in size from 500 to 1,000 pounds. If you have indoor storage space, these can work for you. They stack well, although you'll need a tractor with a loading spike to do it, and they separate into large, easy-to-handle flakes that can be torn apart and fed in mini-size servings.

Nutritious hay is green. Grass hay is usually light to medium green. Alfalfa hay is dark green. Stored hay might be tan or yellowish on the sides that are exposed to the sun, but it should still be green inside. If hay is tan or yellow inside the bale, it means that the hay got sun-bleached from lying in the field too long before it was baled. Although this is not nutritious hay, it's still safe to feed in a pinch.

IS MY GOAT FAT?

A healthy goat with a well-developed rumen looks fat when he's in perfectly good condition. If his belly looks big or he's wide from side to side, that's a good thing. The bigger his rumen, the better he can process his food. Pygmy goats with great rumens look especially fat due to their short legs and blocky body structure.

Feel your goat's rump where his tail meets his body. If you feel blubber, the goat is fat. Also check for excess fat on your goat's lower chest.

Unless they have access to good pasture, goats need year-round access to hay.

Sniff hay before you buy. If it smells dusty, musty, or moldy, even if you don't see obvious mold, don't buy it. Moldy hay can quickly sicken or kill a goat and, unless they're starving, most goats won't touch it. Properly stored, high-quality, year-old hay is a better value than junky current-year's hay. Good hay loses nutritional value rather slowly.

It's always wise to pay for a bale or two and open them to determine their quality before purchasing a large amount of hay. In addition to checking interior color and looking for dried insects, animal parts, and traces of dust or mold, watch for sticks, clods of dirt, and weeds. Although goats like many weeds, some could be poisonous (milkweed and hemlock are often baled into hay in some parts of the country) or have stickers on them that, once dried, can injure a goat's mouth (bull nettle and dried blackberry spring to mind). In any case, you'd be introducing weed seeds that can naturalize and take over your property. It happens all the time.

Storing Hay

Store your hay under cover in a place where sun doesn't beat down on it and compromise its nutritional value. Get stored hay up off the ground. Stack it on old tires, wooden pallets, or poles. Hay in contact with the ground or concrete gets moldy—quickly. Don't let cats use stored hay as their litter box or as a nest to raise their kittens because cats can carry toxoplasmosis. If you buy more hay, rotate older bales to the front and use them first.

When stacking hay, stack the first layer of bales all pointing in the same direction. Then stack the second layer perpendicular to the first layer so that if the first layer of bales is points east–west, the second layer of bales points north–south. Continue alternating layers. This locks the stack in place and makes it more stable.

Feeding Hay

Don't feed hay off the ground. Goats are astoundingly picky eaters, and, if you do, they'll waste more than they eat. There are many useful, ready-made feeders on the market, but it's easy to make your own. We'll talk about that in Chapter 6.

WHEN THERE ISN'T ENOUGH HAY

Although top-quality hay is a goat's best friend, sometimes good hay isn't available, or you don't have a place to store it. That's where hay substitutes come in. Several types of alternative forage worth investigating are bagged, dehydrated hay; Chaffhaye; and Purina's hydration hay blocks (see Resources for suppliers of alternative forage).

Dehydrated hay comes in alfalfa and grass varieties that have been trimmed into short pieces and lightly flavored with molasses.

Chaffhaye is soft, fermented, GMO-free alfalfa hay packed in 50-pound bags. It's great stuff, with one caveat for miniature goat owners: the entire bag must be fed within about a week of opening, or it will mold. If you have enough goats, or you have other animals to feed it to, Chaffhaye is a good choice for alternative forage.

Purina's hydration hay comes in a package of six compact, 2-pound blocks. When added to 5 quarts of water and allowed to set for 10 minutes, each block provides the same amount of forage as a flake of mixed alfalfa and grass hay. Although designed for show horses, it's easily stored and economically feasible to feed as a nutritious, full-time ration to a few miniature goats. It isn't as fibrous as bagged hay, so a small amount of regular, long-stemmed hay should be fed with it.

Folks who keep full-sized goats often supplement their goats' diets with bagged alfalfa pellets. They're readily available at most feed outlets, but they aren't a wise choice for miniature goats because the relatively large size of alfalfa pellets makes them a serious choking hazard. If you feed them, add just enough water to render the pellets into mush before feeding.

Nigerian Dwarf wethers munch hay from a simple fence-line feeder.

Don't feed goats from woven-rope hay nets. Goats get their horns, heads, and legs caught in them. However, fabric hay bags with a single opening on the side work well for a few goats and save a lot of hay.

Be aware that no matter what kind of feeders you use, your goats will waste some hay because once it touches the ground, goats won't touch it. Pick up wasted hay before it's soiled and feed it to a less picky species or use it for bedding in your goats' quarters.

In most cases, it makes sense to keep all-you-can-eat grass hay in front of goats at all times, although you may have to ration if you feed tasty legumes. Feed just enough legume hay to keep your goats in good condition. Avoid feeding legume hay to bucks and wethers because heavy legume consumption contributes to the formation of urinary calculi.

Rumensin: Horse Owners Beware!

Rumensin, a drug added to some bagged goat feed to prevent coccidiosis in ruminants, is very toxic to horses, ponies, donkeys, and mules. A single large serving can kill a large horse. Rumensin is also used in some, but not all, mineral licks labeled for cattle, sheep, and goats, so don't give equines access to mineral licks unless you're sure of the ingredients.

CHOKE

Choke happens when a goat gobbles her feed or tries to eat something that lodges in her throat. A choking goat can still breathe, but she can't swallow. She'll gag, drool, and sling her head around. It's a scary thing, but most chokes resolve themselves in a few minutes. If it doesn't, call your vet without delay.

It's easier to prevent choke than to fix it. A miniature goat's esophagus is small, so don't feed pelleted products designed for full-sized goats, especially if your goat tends to gobble her feed. Instead, choose texturized (granola-type) feeds or reduce regular-sized pellets to mush by adding water before feeding. Chop treats into manageable pieces; for example, don't hand your miniature goat an entire apple or carrot. If she bolts her feed, place large rocks in her feed pan so she'll have to move them around to get to her food, which will slow her down.

Concentrates (Grain)

There are two types of concentrates. The first type is energy feeds that are high in total digestible nutrients (TDN) but low in protein; these include corn, oats, barley, wheat, milo, beet pulp, molasses, and rye. The second type is protein feeds that are at least 15 percent protein, including soybeans and soy meal, cottonseed meal, brewer's grains, and alfalfa pellets. Keep in mind that miniature goats on a maintenance diet do not need concentrates in their diets. Quality hay and goat-specific minerals are sufficient.

You could mix your own grain, but you need a good deal of expertise to do it correctly. It's easier and often better to feed a commercial mix formulated for your goats' needs; there are many options, including organic and GMO-free products. Feed according to the directions on the label, keeping in mind that miniature goats are notoriously easy keepers. If your goats get tubby, cut back.

Goats love to eat paper and cardboard, but more than a bite or two can harm their digestive system.

Minerals

Goats should always have access to goat-specific minerals. Loose minerals are best. Feed them free-choice in a separate feeder or mix in a bit with grain-fed goats' regular rations. Goat minerals served in tubs, also called goat licks, are another option. Because tub minerals are mixed with tasty molasses and served hard in

Utu really gets into his tub minerals!

TUB AND BUCKET CLEANING TIPS

A household scrub brush or long-handled toilet brush and some plain water are all you need to clean buckets and tubs unless there's algae growing in them. To clean an algae-encrusted container, dump it, pour in a generous amount of plain chlorine bleach and then slosh it around or scrub it until all green surfaces are clean.

If you have a bunch of containers to clean, bring them to a central location, pour bleach into the first container, slosh, and then pour the bleach into the next container and keep this up until the bleach is gone. If your goats congregate to watch you work, flip each tub after you bleach it to make sure that no one licks the surface of the bleached tubs (if they can, they will!). When all tubs are bleached, go back to the beginning and thoroughly rinse each container, using a power nozzle on a garden hose. If you have extra tubs that you won't use again right away, flip them over and store them in a convenient place until you need them again.

You can also scrub buckets and tubs with a baking soda paste made by adding a small amount of water to baking soda, which you can buy in bulk at the feed store.

Use a kitchen strainer to remove leaves and other floating debris from your goats' water (or use a secondhand tennis or handball racquet if you find one at a yard sale!). To open a frozen water tank, break the ice with your boot or a heavy object, like an axe, and then scoop out pieces of floating ice with the racquet.

the tub, picky goats that turn their noses up at loose minerals usually love them. Trace-mineral salt blocks are not enough.

Buy a product labeled for goats unless you keep sheep with your goats, in which case you should offer a mineral made for sheep along with a copper supplement.

Copper

Copper is an essential micronutrient that goats must have. In fact, all mammals need copper. Amounts needed vary by species, breed, age, health condition, intake of other minerals, and even levels of feed additives in their diets. Symptoms of copper toxicity in goats include rough coats, anemia, hoof problems, weakness, panting, jaundice, dark red or brown urine, abortion in does, and even death.

A dark-colored coat with an orange overcast is a primary symptom of copper deficiency in goats because copper is essential for the production of melanin, an enzyme involved with hair pigmentation. Low melanin levels cause faded color around the eyes and a red tinge to dark-colored coats. A good mineral formulated specifically for goats usually provides enough copper. When it doesn't, goat owners dose their goats with COWP. COWP stands for *copper oxide wire particles*, which are short pieces of copper wire enclosed in gel capsules and designed to dissolve in a goat's abomasum. These particles slowly disintegrate, releasing copper into the goat's system over a period of time, usually 4 to 6 months. If your goats

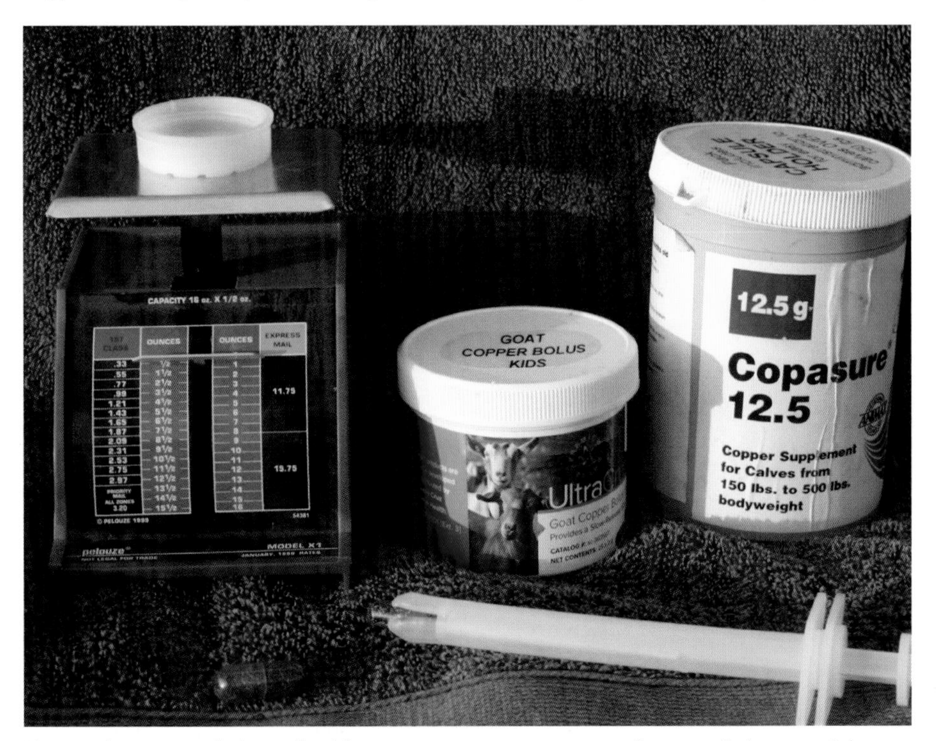

You can buy copper boluses sized for goats, or you can cut open Copasure boluses, weigh their contents, and repack the copper rods into goat-size gelatin capsules.

GARDENING FOR GOATS

If you enjoy growing vegetables, consider gardening for your goats. Goats can eat modest quantities of almost every garden goodie except onions, raw potatoes, and vines from nightshade plants like eggplants, tomatoes, and potatoes. They love lettuce, turnip greens, collards, chard, melons, summer and winter squash, pumpkins, peas, kale, corn, and most kinds of garden refuse. Winter squash and pumpkins are best bets for goats because they're nutritious and easy to grow, and they store for months under the right conditions.

If you want to plant something just for your goats to eat, try mangels, a.k.a. mangel-wurzels or mangolds, which were once a staple of farmyard animals' winter diets. A mangel is a type of huge beet with red, yellow, or white flesh. At maturity, they weigh from 10 to 20 pounds each and are up to 2 feet long. They're easy to grow and yield a huge crop; heirloom gardening catalogs carry mangel seed.

After you harvest the mature mangels, age them in storage until January and then chop them in small pieces as needed and feed them raw as supplementary fodder. A dozen mangels can keep a few miniature goats happy for months. Store mangels in a root cellar or in an alternative storage container, such as a recycled refrigerator or a dirt-filled garbage can. They stay fresh in storage for about 6 months.

Goats that can't get out to select their own forage appreciate cut-and-carry greenery. The trick is making sure that what you bring them isn't poisonous to goats and that you feed it in moderation. Scope out the poisonous-plants resources at the end of this book, and if you can't identify a tree or plant, take a sample to your cooperative extension agent for positive identification.

show signs of copper deficiency, and you want to supplement them with COWP, it's best to buy copper boluses sized for goats. They're generally dosed at the rate of 1 gram of COWP per 22 pounds of goat.

Copper boluses are especially important when people keep sheep and goats together. Sheep need copper but usually get enough through their diets. Sheep that consume bagged feed or mineral products formulated for goats retain excess copper in their livers. This builds up and eventually kills them, so if you have both sheep and goats, you need to do one of several things.

- Goats are good climbers, but most sheep aren't (some hair sheep breeds are the exception), so it's usually safe to put goat minerals where goats can hop up on something to eat them but sheep can't. The disadvantage is that sheep need sheep minerals down at their own level, and if goats eat that instead of their goat minerals, they might not ingest enough copper.

- Separate your sheep at night and provide species-specific minerals in each group's sleeping area.
- Put out sheep minerals for everyone and copper-bolus your goats two or three times a year.

There are several ways to dose your goats with COWP. The best is to use a pet piller or a balling gun (see Chapter 8) to deposit a properly sized and packaged copper bolus on the back of your goat's tongue. This delivers copper wires to your goat's abomasum intact, which is as it should be. Some goats are notoriously hard to dose, so some goat owners cut a marshmallow in half, squash it a bit, scoop out a pocket, and insert the wires from a single bolus into the marshmallow, and they feed the wires to their goats in this manner.

Salt

Most mineral mixes contain salt, but it's best to feed additional salt on the side. Granular salt can be served in a loose mineral feeder or as salt blocks. If you choose blocks, place them in containers so that they're up off the ground; otherwise, the bottom of the blocks will eventually melt.

Salt blocks come in plain (white), sulfured (yellow), trace mineral (brown), red (iodine added), and blue (cobalt and iodine added) varieties. Specialty salts like Redmond salt, a premium natural salt mined in Utah, and sea salt sometimes come in spools that you can hang in your goat shed. In most cases, plain white salt is sufficient.

Mom and kid graze together.

Water

Miniature goats drink between a quart to a gallon, or even more, of water per day. Lactating does have the highest water requirements. To help prevent urinary calculi, it's important for wethers and bucks to drink a lot of water, too.

If your goats' drinking water is contaminated with droppings, algae, dead bugs, leaves, or other debris, they'll drink just enough to get by. Goats are picky. If you wouldn't drink it, your goats probably won't drink it, either.

Don't use flimsy household plastic pails around goats because they'll trash cheap buckets in record time. Food-service buckets and heavy-duty plastic buckets and troughs designed specifically for livestock are a better bet.

Place water-filled tubs and buckets in the shade during the summer months to help inhibit algae growth—the fresher the water, the more your goats will drink. When it's very hot outside, freeze ice in plastic milk bottles and place one in each trough or tub of water. Refreeze the bottles overnight, and they'll be ready to use again the next day.

During the winter, keep water supplies from freezing by installing bucket heaters. A word of caution: encase the cords in PVC pipe or garden hose split down the side and taped back together with duct tape. If you don't, your goats could gnaw through the cord and electrocute themselves. Another cold-weather watering option is to carry buckets of warm water out to your goats several times a day, bringing frozen buckets indoors to thaw out.

Another word of caution: be careful about water containers in kidding stalls or places where young, active kids like to play. Does have been known to deliver kids into buckets, where they drown, and it's easy for a small kid to leap into a bucket of water that is deeper than he is tall. Shallow plastic feed and water pans designed for large dogs work well in both instances. It's better to set out several shallow pans than one deep pail.

GOAT KEEPER'S NOTEBOOK

Shop Around
If you feed pre-bagged grain mixes, shop around when buying your feed. Small mills and local manufacturers often price quality products much lower than the big-name mills.

Store It Safely
Store feed where your goats can't break in and overeat. Appliance shops sell or give away broken chest-type freezers that are perfect for storing grain, but be sure to remove the freezer's locking mechanism so children can't get trapped inside. Another option is to invest in goat-proof locks for feed-room doors. Use at least two on each door. Goats are smart!

Weigh It
Measure grain by the pound instead of by the gallon. If you don't have a farm scale to weigh your hay and grain, an accurate bathroom scale will do. To weigh heavy feed with a bathroom scale, place the scale on a firm, flat surface. Weigh yourself first and then reweigh yourself holding a quantity of feed and subtract the difference.

Coffee Cans and Ice-Cream Pails
Weigh each milking doe's grain ration and place it in a lidded coffee can or small plastic ice-cream pail with her name printed on top so it's handy when she gets up on the milking stand.

Save the Strings
If you buy small square bales tied with plastic twine, save the twine for other uses. Cut bales where the twine is tied, trim off the knot, and then tie the strings together end-to-end and roll it into a ball. This is easiest if you start by wrapping the twine around a short length of stick to help form the ball. Twine comes in handy in so many ways, and plastic baler twine is strong.

Carrying Water
Recycle plastic milk jugs for carrying hot water or for taking water from home along to a show. Jugs of water are easily transported by hand or in a garden cart, van, or truck. To prevent sloshing and spilling when transporting larger amounts of water, use a clean, covered 5-gallon food-service bucket or a trash can with a secure lid. Or line a lidless container with a strong trash bag, fill the bag with water, and then tie the bag shut.

Water Away from Home
If you don't want to carry water when traveling, begin placing a small amount of flavoring, such as Tang or Kool-Aid, in your goats' drinking water a week or two before the trip. Then add the same flavoring to any water encountered in your journey, and it will make the strange water taste like home.

Frozen Faucets
If your outdoor water faucet freezes, thaw it quickly with a handheld hair dryer.

CHAPTER 6

Shelter and Fences

You don't need to build a palace to house your mini goats. Goats need sufficient room in a draft-free, well-ventilated structure, where they can get away from bad weather and nasty flies, along with an exercise area or pasture to stretch their legs. If you can provide that, you will have happy goats.

How much room? That depends. If you live in a fairly mild climate, and your goats have room to roam outdoors, plan on at least 10 square feet of communal indoor space for a tiny Nigerian Dwarf or Pygmy Goat and 16 to 18 square feet for mid-sized breeds like Pygoras and Kinders. Plan for extra space in cold climates or other situations in which your goats will spend a lot of time indoors.

Experts disagree on how much outdoor exercise space goats ideally need. It takes tall, sturdy fences to keep goats, even mini goats, inside their enclosures while keeping goat-killing predators out, so your goats are safer in a well-fenced small exercise yard than they are roaming in a poorly fenced large pasture. A big area is naturally better, but if you add several types of climbing apparatus and some toys to the equation, goats can get considerable exercise in a smallish space. Figure at least 25 square feet of outdoor space for a tiny mini and 30 to 35 square feet for each mid-sized goat. Keep in mind that overcrowding stresses goats, and stress leads to increased levels of aggressive behavior. You won't like it. Give your goats as much space, indoors and out, as you can.

Goat Shelters

If you'd like to build your goats their own fancy barn, they'll love it—but miniature goats can be happily housed in refurbished chicken coops or similar farm outbuildings, garages, Southern-style carports with the open sides closed in, homemade hoop houses, goat-sized field shelters, prefab garden sheds, plastic or fiberglass calf hutches, Quonset-style portable

A STALL IN THE BARN

If you own a horse barn with stalls, a spare stall makes a perfect home for three or four minis, depending on their size. Install a manger, a water-bucket hanger, and mineral cups on their level, and a standard-size horse stall will be first-class digs for your mini goats.

If you'd prefer to build individual stalls for each goat, make each stall at least 4 by 6 feet in size. You'll have to get in there to clean, and it's hard to wield a pitchfork or broom in a stall much smaller than that.

No matter where you build or use existing stalls—in a barn, garage, or other type of outbuilding—be sure that the goats living in them can see other goats. Sheep panels—a type of prefabricated, heavy-duty mesh fence made with galvanized ¼-inch steel rods—are 48 inches tall and 16 feet long, with closely spaced bottom stays. When trimmed to size, these make excellent inexpensive stall partitions if set up where drafts aren't a problem.

A large dog house makes a fine paddock shelter for miniature goats.

huts designed for pigs, and livestock sheds crafted out of freebie pallets. Some of these structures require modifications to render them suitable as goat housing, but that's doable.

Exercise and pasture shelters are easier still. An extra-large dog house makes a fine shelter for several kids or a pair of Pygmy Goats. Children's playhouses work quite well, as do pickup-truck bed toppers elevated a bit off the ground. Think outside the box. If it's big enough and it gets goats out of the weather and up off the ground, it could work.

Building for Goats

Your goats' primary shelter should be built or installed on an easily accessible, well-drained site. Electricity and running water are bonuses. For your sake, your goat structure should be tall enough for you to stand up in, or it should be easily movable; otherwise, cleaning it is going to be a chore.

Don't use particle board or pressed board. These flimsy manufactured wood products are made using formaldehyde, a known carcinogen and an eye and respiratory-tract irritant. Worse, particle board and pressed board are candy to most goats. They'll eat holes in it, making you mad and making them sick. Build with real wood when you can. When you can't, choose heavyweight exterior plywood that goats aren't likely to quickly demolish.

Goat housing must be adequately ventilated, even during the harshest winter months. Goats tolerate a lot of cold if they're dry and can stay out of drafts. Don't close your building up tight; a lack of ventilation can lead to respiratory distress and pneumonia.

Avoid dangerous heat lamps. Curious goats have been known to pull heat lamps down into their bedding and set their barns on fire. For the same reason, as well as to avoid mouth injuries, cover electrical wiring or electrical cords that goats can reach with PVC pipe or with rubber or vinyl hose cut down one side and resealed with duct tape.

Multilevel luxury goat accommodations designed by Don Kneass of the Goat Justice League.

Field Shelters

A field shelter, also called a run-in shed or loafing shed, is a four-sided structure with three enclosed sides, a fourth side open to the elements, and a slanted roof. Field shelters are easy to build, are nicely ventilated, and can be used year-round unless you live where bitter cold and snowy winters are the norm. Erect a field shelter in a well-drained

THE PERFECT URBAN GOAT BARN

If you want to build a super-efficient, attractive combination-goat-and-chicken shelter, consider building the Goat Justice League's state-of-the-art goat shed. Designed by Don Kneass, the husband of Seattle author Jennie Grant, and featured in *City Goats: The Goat Justice League's Guide to Backyard Goat Keeping* (see Resources), it offers comfortable, multiple-level housing for a pair of dairy goats and four laying hens along with feed storage areas and room for a milking stand and stool. Jennie says, "While keeping the goats warm and dry, it doubles as a play structure, increases the goats' living area with a rooftop deck, and includes a chicken coop, a hay storage box, and a small shed to store chicken and goat feed." To learn more, check out *City Goats*. It is a fun and informative read, especially for urban goatkeepers.

GOAT BLANKETS AND COATS

If you think that your goats are cold, don't close the doors and windows—instead, dress your goats! There are wonderful goat coats on the market like this beautifully fitted goat blanket from Horsewear Ireland (see Resources). Hoegger Supply and Caprine Supply sell goat blankets, too.

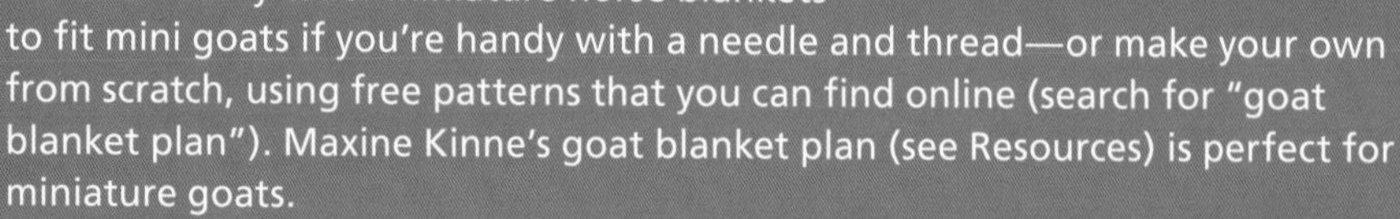

You can easily tailor miniature horse blankets to fit mini goats if you're handy with a needle and thread—or make your own from scratch, using free patterns that you can find online (search for "goat blanket plan"). Maxine Kinne's goat blanket plan (see Resources) is perfect for miniature goats.

One of the quickest and least expensive ways to clothe your goats is to scope out yard sales or visit your favorite charity's thrift shop and pick up some cardigan sweaters; wool sweaters are best. To use them, cut the sleeves off a few inches from the body of the sweater, thread your goat's front legs through the openings (with the back of the sweater to her belly), and then button the sweater along her back. Better yet, buy oversized wool sweaters and shrink them by machine washing them in warm water with plenty of soap. Felted sweaters don't unravel, and they're toasty warm.

Another ploy is to try heavy-duty coats and wool sweaters made for dogs. Small ones fit kids, and extra-large sizes fit small adult minis.

Two caveats: When adapting a garment not originally made for goats, make sure that it doesn't cover a wether's penis. Also, if a garment seems to slip around too much (you don't want your goat to get her legs tangled in it), stabilize it by adding elastic leg straps like the ones in this picture.

location, with the open side opposite your prevailing winds (oriented to the south in most locations) and the roof slanted to the back so that rain and snow will not slide off in front of the shed. Field shelters can be easily converted into covered pens with the addition of a farm gate across the open side.

Ready-Made Housing

If you aren't handy with tools but want a safe, secure shelter in a hurry, investigate Quonset-style portable huts or plastic or fiberglass calf hutches.

Portable Huts

Portable huts are built by various companies, including Port-a-Hut, Inc. in Storm Lake, Iowa (see Resources), and sold by retailers throughout most of the United States. They come in various sizes. Port-a-Hut's structures come in six sizes, and their most popular model is a 4½-foot wide, 7½-foot long, 46-inch tall structure designed to house pigs. One of these huts can comfortably house five or six Nigerian Dwarfs or Pygmy Goats and up to four Pygoras or Kinders. Several basic Port-a-Huts (they come fully assembled) can be nested and hauled in the back of a pickup truck and moved from place to place by two strong people. In between moves, the huts must be fastened to the ground; otherwise, strong winds can bowl them over. Port-a-Hut provides three pound-in-the-ground-type anchors with each hut. They're almost indestructible and sized so that larger minis can leap up on top and play "king of the mountain." The only downside is that it's hard for an adult to get inside to clean them.

Calf Hutches

Calf hutches are designed to house young dairy calves, and they come in several styles and sizes. Even the smallest is roomy enough for a pair of mini goats, and because they look a lot like giant doghouses, they fit into urban and suburban settings quite well. Calf hutches are cooler in the summer and warmer in the winter than steel portable huts; some units have doors, and the smaller sizes come fully assembled. Find new calf hutches at farm stores and used ones via Craigslist or from local dairymen.

BUILD IT WITH PALLETS

Building with pallets is "in." If you search for "pallet barn" or "pallet shed" on Google or YouTube, for example, you'll find all the inspiration you need to build sturdy, inexpensive shelters for any number of goats. You can also use pallets to craft gates, fences, feeders, and just about any other barnyard structure made of wood.

Wooden pallets offer a great combination of weight, stiffness, durability, and cost (they're usually free). They come in single- or double-face types, and some are reversible, meaning that either side can be used for the top. Most of the pallets made in North America are made of oak or southern pine and measure 40 by 48 inches in size. Nowadays, pallets require a logo that tells how the wood was treated. Because many goats are wood nibblers, avoid pallets marked HT (heat-treated, possibly with chemicals) or MB (treated with methyl bromide, a neurotoxin and carcinogen). Also avoid pallets used by pesticide or chemical companies due to the risk of spills that could have impregnated the pallets with toxins.

Small businesses, such as hardware, automotive, garden, farm, and feed stores, usually give pallets away. Or ask at Freecycle (www.freecycle.org), and you might find someone offering a pile of pallets for free.

Goats enjoy the comforts of home, too!

Bedding

The floors of goat shelters can be made of dirt, sand, packed clay, gravel, wood, or concrete. They must be liberally topped with straw, wood shavings, peat moss, shredded paper (banks generate a lot of this and usually give it away for free), chipped corn cobs, rice, or peanut hulls to keep your goats warm in the winter, comfortable, and dry.

The bedding you choose should be readily available; safe for goats to nibble on; absorbent; easy to clean, handle, and store; not dusty; and cost effective. If you plan to compost it, it should also compost well.

Alternatively, you might choose a bare concrete floor with raised sleeping platforms. Another option is a layer of rubber horse-stall mats over dirt, packed clay, sand, gravel, or concrete with little or no bedding on top of the stall mats. If you go this route, you

Goat-Bed Tips

Make comfortable platform beds for your goats by nailing sheets of sturdy plywood to freebie wooden pallets.

Exercise rebounders and elevated dog beds also make great goat bunks. Just make sure that any openings around the edges are not large enough not to catch and hold a goat's leg if one slips through.

For larger groups, set up a full-size trampoline where your goats can lie on it in their yard or barn. Provide boxes for them to use as steps to reach it.

CART IT AWAY

An essential tool for a goat keeper is a wheelbarrow or utility cart for toting manure from the goat shed to the muck heap and for moving heavy items like bales of hay. You might find a used wheelbarrow at a yard sale or farm auction, but if you're going to buy new, a utility cart—also known as a garden cart or feed cart—is a much better choice. This is a nicely balanced box-like wooden or plastic contraption with big, bicycle-style wheels that make it is easier to maneuver and push than a wheelbarrow. Besides, goats love to play and snooze in utility carts. Once you try one, it will become the one piece of goat equipment that you'll never be without.

must provide your goats with raised beds for warmth and so they don't have to lie in urine; concrete floors may be too cold for northern climates. The beauty of a bare floor is that you can sweep it every day, so it's easy to keep clean. Concrete has an additional advantage: it helps wear down goats' hooves, so they needn't be trimmed as often.

Exercise Yards

An exercise yard can be constructed on a concrete pad, natural dirt, or a textured surface such as sand or gravel. Make it interesting. Build a simple playground so that your goats stay amused.

Apart from adequate size, the most important part of an outdoor exercise yard or goat pasture is the fencing used to contain it. Consider this: according to the United States Department of Agriculture's National Agriculture Statistics Service's Sheep and Goat Loss Survey, conducted in 2010, 60,000 adult goats

FENCING TIPS

The hows and whys of erecting pasture fencing is beyond the scope of this book. For inspiration and instruction, talk with your cooperative extension agent, read one of the fencing books listed in the Resources, or check out the fencing section of Premier1 Supplies' website (www.premier1supplies). Premier1 also offers a free fencing catalog that is packed with must-read how-to information.

and 120,000 kids were killed by predators in 2009 alone. Goats are defenseless against predators, especially goats contained in small enclosures, so exercise-yard fences must be extra sturdy and at least 4½ feet tall. You don't want dogs to get in and hurt or kill your goats. Yes, dogs. Coyotes are a problem in rural areas, as are large animals like bears and cougars in some parts of the country, but free-roaming dogs are everywhere, even in urban and suburban settings, and predation by dogs isn't pretty. A single dog can chase down and fatally maim a goat in record time. In addition to building tall, strong fences, if you live where dogs are a nuisance, keep your goats indoors at night, as well as during the day when no one is home to protect them.

A good way to fence an exercise yard is with cattle panels set on wooden fence posts. Cattle panels, sometimes called stock panels, are the big brothers of the sheep panels we mentioned earlier in this chapter. Most cattle panels are 52 inches tall and built using 8-inch stays; horizontal wires are set closer together near the bottom of the panel to prevent kids from escaping. Cattle panels are usually sold in 16-foot lengths that can be easily trimmed to size using bolt cutters. A bad thing about sheep and cattle panels is that the raw end of each rod is very sharp. To make them more user-friendly, smooth each rod end with a rasp or tap its sharp edges with a hammer to flatten it out.

Pasture Fencing

Goat-friendly fences are usually built with stout woven wire, but when price is no object, cattle panels work even better. What doesn't work is barbed wire or most electric fences (some electric fences can be OK with careful planning and many strands of wire). Portable net fencing is good in some situations, but it isn't suitable for horned goats or mini kids, who can get their heads caught in the netting and electrocute themselves. It can't be said too often: a properly built exercise yard is a safer and better bet for miniature goats than a pasture surrounded with flimsy or poorly constructed fencing.

Chew on This

Add enough liquid soap to ground cayenne pepper to make a thin paste and paint it on surfaces on which your goats like to chew. Or mix ground cayenne pepper with petroleum jelly to make a smear-on paste. Plain bar soap rubbed on surfaces works, too.

Another option is to cut down saplings and bring them to your goats. They'll eat the leaves and gnaw on the branches and leave your ornamental trees and wooden fences alone.

Organize It

Organize your goat-keeping equipment so you can find an item when you need it. If you can't find it, you'll waste good money buying another one. Hang items like halters, leads, and collars neatly on the wall. Mount a wooden pallet on the wall, with the open ends at the top and bottom, and slip rakes, brooms, pitchforks, and the like in between its decks.

A Goat Keeper's Toolbox

Goats are destructive creatures, so plan to make ongoing repairs to fences, shelters, and almost anything else your goats interact with. Buy a sturdy toolbox and stock it with duct tape, wire ties, bolt cutters (needed to quickly cut cattle panels and woven wire fencing in an emergency), pliers, a screwdriver, a sturdy hammer, at least one pair of heavy leather gloves, and an assortment of nails, screws, and fence staples. Keep an eye out at yard sales and farm auctions, where you may have the opportunity to buy older, better quality tools at decent prices.

Keep your tools stowed in their box and the toolbox in a designated spot where anyone can access it quickly in an emergency.

Are They Breathing Fresh Air?

Ammonia buildup in goat stalls and shelters leads to serious respiratory problems, like pneumonia. To check if ventilation and cleanliness are adequate, don't just walk into the stall and sniff the air; instead, get down on your hands and knees and sniff about 8 inches from the ground—that's mini-goat breathing height.

CHAPTER 7

Parasites

Resistance to goat anthelmintics (worming drugs) is the most serious problem facing goat owners today, not just here in the United States, but around the world as well. Goats are extremely prone to parasitism. Many goats die of worm infestation every year, so it's important to keep your goats as parasite-free as you can. Your veterinarian is your number-one source of information about the internal parasites prevalent in your area. He can work with you to create an ideal worming regimen for your goats. Talk to your vet. Your goats will thank you.

Be in the Know

To find out how many and what type(s) of parasites your goats are carrying, your vet will run eggs per gram (EPG) fecal tests. To do this, he'll need fresh manure from each goat or, if you have a lot of goats, from at least five to ten goats selected at random. The key word is *fresh*. Stand by with a labeled plastic sandwich bag turned inside-out on your hand, and, when each goat delivers, pick up a sample, turn the bag right side out around the fecal material, and seal it up. Samples don't have to be refrigerated.

Most internal parasites are too small to see with the naked eye, so after processing the manure, your vet will look at it under a microscope to identify what types of parasite eggs are in there. Based on the types of parasite eggs and the number of each type of egg found in 1 gram of prepared sample, he can recommend the best worming agents for your goats.

Wormer Resistance

In the past, experts told sheep and goat producers that to effectively control worms in their flock, they should worm all of their animals at the same time and should rotate wormers to reduce drug resistance. Since most people don't weigh their stock before

Sheep Are the Same—Sort Of

Sheep and goats share many of the same illnesses, worms, and external parasites, but the amount of wormer needed to effectively dose goats and sheep isn't the same. In most cases, goats require considerably more product than is used to worm sheep, so don't dose your goats based on the sheep dosage on product labels.

worming, many were underdosing their animals. In this manner, sheep and goats were exposed to all of the available worming chemicals, often in doses too small to kill all of their worms. Weaker worms died while the strong survived, and eventually wormer-resistant superworms evolved to the point that wormers lost their punch—and drug companies aren't developing new products to replace the old ones.

It's important to have a veterinarian run fecals on your goats. If you have a good microscope, you can learn to run them yourself. Don't worm your goats indiscriminately because that leads to further wormer resistance. And if a product isn't working on your farm, you need to know it.

Some things you can do to help keep wormer-resistant parasites from moving in on your goats include:

- Don't rotate wormers every time you dose your goats. If a wormer is working, use it for at least a year or until it loses its punch.

If left unchecked, worms can spread quickly where multiple goats are kept together.

ROUNDWORM REPRODUCTION

The most dangerous internal parasites of goats—barber pole worms and brown stomach worms—are both roundworms. Heavy loads of either type of worm can easily kill a goat. Here's what you need to know about roundworms:

- Roundworms are a single-host parasite; they live and reproduce in a single goat.
- Adult barber pole worms and brown stomach worms live in the abomasum compartment of the goat's stomach, where they feed on red blood cells and lay eggs.
- Eggs are passed out in manure but are too small to be seen with the naked eye.
- Eggs hatch and then progress from first-stage to second-stage and then third-stage larvae. As third-stage larvae, they crawl up on blades of grass and wait for a host to happen along.
- When a goat nibbles the grass, she also ingests the larvae.
- The larvae mature into adult worms in the intestines and stomach, they lay eggs, and the cycle begins again.

- Worm only the goats that need it, not every goat you own, based on fecal exams or checking your goats' inner eyelids for signs of anemia. According to research conducted at Langston University in Langston, Oklahoma, only 20 to 30 percent of the goats in most herds carry most of the worms and shed the most eggs. These goats require frequent worming. The rest don't.
- Ten days after worming your goats, collect more samples and run new tests. If your worming program is working, there should be an 80- to 95-percent reduction in the number of eggs per gram of manure.
- Weigh your goats, and make sure that you aren't underdosing. Goats aren't dosed with the same amount of wormer per pound as sheep, horses, or cattle. Ask your vet for the correct dose or consult one of the dosing charts listed in the Resources at the back of this book.
- With few exceptions, wormers should be delivered by mouth. No matter what the labels say, don't inject them, and never use them as pour-ons (pour-ons formulated for cattle can damage a goat's more delicate skin). Use a dosing syringe to deposit the wormer on the back of the goat's tongue and then elevate her chin until she swallows. We'll show you how in Chapter 8.
- Don't use pour-on products orally, even though many people do. Pour-ons contain petroleum products that can burn a goat's mouth and throat. We know this firsthand. Trust me—don't do it.
- Immediately worm and then quarantine new goats.
- Worm does soon after kidding, when fatigue and stress lower their resistance and changing estrogen levels cause worms to proliferate.
- Since worm larvae tend to stay within 3 inches of the ground, move your goats to new pasture before grass is overgrazed.

HOW MUCH DOES YOUR GOAT WEIGH?

To properly dose some pharmaceuticals and all wormers, you need to know how much your goat weighs. The best way to find out is to weigh him on a scale. If he's a kid, or if you're strong and he's a tiny adult mini, pick him up and weigh yourself on a bathroom scale. Put him down and weigh yourself again, this time without holding the goat, and subtract the difference.

Large-animal veterinarians have livestock scales, and small-animal veterinarians usually have walk-on scales at their clinics. Call and see if you can weigh your goats at one of those facilities.

You can tape-weigh your goat using a dressmaker's cloth measuring tape. Restrain the goat so that he's standing on a level surface with all four feet placed squarely under him. Place the tape around his body, just behind his front legs at a point slightly behind his shoulder blades, making sure that the tape is perfectly flat and not twisted. Draw the tape up all the way to the skin until it's snug but not tight. This is his heart girth measurement.

Next, measure the length of his body from the point of his shoulder to his pin bone (the pointy bone near his tail). Multiply his heart girth by his body length and divide by 300 to find his weight in pounds (heart girth × heart girth × body length ÷ 300 = weight in pounds). Alternatively, you can purchase a goat-weight measuring tape that does the calculations for you.

Tape-weighing works for proportionally sized goats like Nigerian Dwarfs, miniature dairy goats, and Miniature Silky Fainting Goats but not so well for broad, short-legged breeds like Pygmys, Pygoras, and some Kinders. For those breeds, consult Kinne's Minis' Pygmy-Goat measuring chart (see Resources).

Barber Pole Worms

Barber pole worms (*Haemonchus contortus*) are the most serious internal parasite of goats in North America. They are ¾ to 1 inch long and tapered at both ends. Females are red-and-white striped like an old-fashioned barber pole, and males are solid red. They live in a goat's abomasum or true stomach. They're a worldwide threat but are especially troublesome in hot, wet climates.

Goats consume barber pole worm larvae while grazing. Kids become infected when they start to eat grass. Once inside a goat, the larvae of barber pole worms burrow into

Kids risk barber pole worm infection when they start eating grass.

the lining of their host's abomasum, where they feed on red blood cells. They molt twice before becoming adult barber pole worms. Female worms lay from 5,000 to 10,000 eggs every day, which the goat passes in his manure out into the pasture, where eggs hatch and the cycle begins again.

Adult barber pole worms also feed on their host's blood, so goats with heavy worm loads quickly become anemic. Signs of barber pole worm infestation include diarrhea, dehydration, rough hair coats, incoordination, lethargy, bottle jaw, and pale mucus membranes. Some goats die as a result of the infestation.

Bottle jaw, also called mandibular edema, happens when fluid accumulates under a goat's jaw. Fluid also builds up in the animal's abdomen and gut wall, but you can't see that with your naked eye. If your goat gets bottle jaw, he is very close to death. However, don't mistake milk goiter, a benign swelling that some kids develop, for bottle jaw. Bottle jaw is puffy and hangs down from the jaw itself, whereas milk goiter occurs at the juncture of the lower jaw and throat.

Swelling directly under the jaw, pictured here, is indicative of bottle jaw.

A good way to check your goats for barber pole worm infestation is to examine the membranes around their eyes at least every other week during the hot, damp summer months and monthly throughout the rest of the year. South African livestock specialists

THE EYES HAVE IT

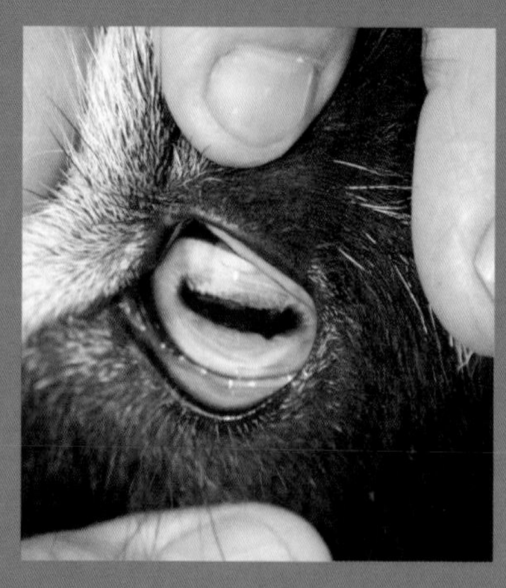

The thin skin that lines the inner surface of the body is called the mucous membrane. The surfaces of mucous membranes are good indicators of what's happening inside the body because they're so thin and transparent that you can see blood vessels through them. The easiest place to examine mucous membranes is inside the eyelid.

Check the inner eyelid color in natural light. To open the eye, push the upper eyelid up with one thumb while pulling the lower lid down with the other thumb. Look at the color inside the lower eyelid, keeping the eye open for just a very short time.

Healthy goats have dark pink to reddish mucous membranes. Pale pink or white membranes indicate anemia, usually caused by a heavy worm load. If the membranes are white, the goat is in serious trouble and needs to be wormed right away.

developed a diagnostic procedure called FAMACHA (see Resources) that you can learn by attending a FAMACHA clinic, but you can also run a basic test without FAMACHA training.

Brown Stomach Worms

The brown stomach worm (*Telodorsagia circumcincta*) is smaller and less common than the barber pole worm. It's dangerous because it feeds on nutrients harvested from the stomach's mucous lining and can permanently damage a goat's abomasum, affecting her ability to digest nutrients. Symptoms include watery green diarrhea, rapid weight loss, rough hair coat, and mild to moderate anemia. Brown stomach worms are a cool-season parasite. Hot, dry conditions kill the larvae, but new eggs hatch in autumn, and the cycle starts again.

Those Other Worms

Only a fecal exam can tell for sure whether your goats are infested with the following worms.

Liver Flukes

The liver fluke (*Fasciola hepatica*) requires an intermediate host—a snail—to reach infection stage, so they're only a problem where goats graze wetlands, especially in hot, humid locales. Once inside the goat, liver flukes feed on bile-duct linings and can cause scarring and cirrhosis of the liver, possibly leading to death. Symptoms of liver fluke infestation include anemia, extreme weight loss, and low milk yields.

Bankrupt Worms

Bankrupt worms (*Trichostrongylus colubriformis*) and liver flukes thrive under the same conditions: wetlands, heat, and humidity. Like brown stomach worms, bankrupt worms interfere with their host's digestive capability, causing diarrhea, rough hair coat, and weight loss.

Lungworms

Goats become infected with lungworms (*Muellerius capillaris*) when they consume lungworm larvae in manure. Once ingested, larvae travel to the lungs, where they can cause respiratory problems. There usually aren't any obvious symptoms of lungworms except in extreme cases, when goats cough and wheeze. Wormers that control barber pole worms usually control lungworms, too.

Tapeworms

If your goats have tapeworms (*Moniezia spp.*), you'll probably see shed segments resembling white rice in their manure. A heavy load of tapeworms slows growth in kids, but unless an intestinal blockage occurs, tapeworms aren't a major problem.

Meningeal Worms

Meningeal worms (*Parelaphostrongylus tenuis*) are sometimes called deerworms or meningeal deerworms because their natural host is the white-tailed deer. Goats (as well as sheep, llamas, alpacas, and wild species, such as moose) are at risk wherever white-tailed deer are present. Ground-dwelling slugs and snails are the intermediate host between deer and other species.

Meningeal worms don't seem to bother their natural host very much, but, in other species, ingested larvae migrate to their host's spinal cord and brain, which causes weakness in the back legs, a staggering gait, circling, gradual weight loss, and paralysis, leading to death.

In places where meningeal worms are a problem, some vets prescribe preventative ivermectin injections at 30-day intervals throughout the spring and summer

Meningeal worms are a risk for goats in areas where white-tailed deer (pictured) are present.

months. Treatment is difficult and often fatal. If you live where white-tailed deer are present, discuss meningeal worms with your veterinarian. Meningeal worm infestation in goats is on the rise.

Wormers 101

Very few wormers are labeled for goats, so the products you'll use to worm your goats are considered "off-label." Technically, you can obtain them only through a prescription from a veterinarian. In fact, they're readily available by mail order and off the shelf at farm stores and saddlery stores. Three classes of drugs are used to worm goats:

Benzimidazoles: fenbendazole, albendazole, oxybendazole, and thiabendazole

Nicotinics: levamisole and pyrantel

Macrolytic lactones: ivermectin, doramectin, and moxidectin

Benzimidazoles, marketed as Safeguard, Panacur, Valbazen, and Synanthic, are called white wormers. They are broad spectrum, safe to use, and effective against tapeworms. Albendazole is also effective against adult liver flukes, but should not be used to worm pregnant or lactating does.

Levamisole, also marketed as Tramisol, is a clear wormer. It's a broad-spectrum product, but it has a narrower margin of safety than most other wormers and should never be given to pregnant does. Pyrantel, marketed as Strongid, is labeled specifically for goats and is effective only against adult worms. Unfortunately, nowadays worms are largely resistant to pyrantel.

The macrolytic lactones, also known as avermectins and marketed as generic ivermectin products, Ivomec, Dectomax, Quest, and Cydectin, are the newest family of drugs. They are broad-spectrum wormers with a wide margin of safety.

There are also herbal wormers on the market, although the jury is out regarding their effectiveness. Some goat owners love them; others say that they just don't work. They're worth a try if you raise your goats organically (or attempt to), but you will need to monitor your goats for worm infestation: have fecals run and check their mucous membranes on a regular basis. If you aren't happy with the results, switch to a chemical wormer.

External Parasites

External parasites include lice, mites, ticks, and biting flies that feed on blood, skin, and hair.

Lice

Goat lice are host-specific and only infest goats and sheep. Both immature and adult lice suck blood and feed on skin. Lice are usually spread via direct contact. They reach peak numbers in late winter and early spring. Summer infestations are fairly rare.

Lousy goats have dull, matted coats, and they spend a lot of time scratching and grooming themselves. Scratching leads to bald spots and raw patches on the skin. They may not eat, so they'll lose weight and give less milk.

COCCIDIA KILL

Coccidiosis, often called cocci, is a common, potentially fatal, and highly contagious disease of goats, especially young kids. Other livestock species get coccidiosis, too, but Eimeria protozoa, the single-cell parasites that cause it, are species-specific, so your goats can't catch dog coccidiosis or chicken coccidiosis and vice versa.

Coccidia are everywhere. All goats are infected to some degree. A single badly infected goat can shed thousands of microscopic oocysts (the protozoan equivalent of eggs) in her droppings every day. If another goat, especially a young kid, ingests a mature oocyst, she can become ill a week or two later. As oocysts multiply and take over the gut, they destroy their host's intestinal lining. Without immediate, aggressive treatment, kids either die or develop chronic coccidiosis that leads to badly stunted growth.

Without preventative treatment, most kids raised around other goats come down with coccidiosis at around 3 weeks of age. Symptoms include watery diarrhea, sometimes containing blood or mucus; dehydration; listlessness; poor appetite; and abdominal pain. Products that kill worms have no effect on coccidia, so standard worming protocols won't prevent or treat coccidiosis. Before your first doe kids, discuss cocci prevention with your vet.

External parasites are often spread through contact, and some can affect both goats and sheep.

There are two types of lice: sucking lice, which pierce their host's skin and draw blood, and biting lice. Loss of blood to sucking lice can lead to serious anemia. The following three types of sucking lice infest goats:

- *Foot lice* prefer the feet and legs of sheep and goats. They sometimes infest the belly area, too, and scrotum infestations on bucks are relatively common. Kids tend to have the highest rate of infestations.
- *African blue lice* are a problem in wet, warm parts of the United States. They're usually found on the goat's body, head, and neck. Heavy populations can kill a goat.
- *Goat sucking lice* feed all over the animal's body. They can infest sheep, too.

Biting lice have chewing mouthparts, and they dine on scabs, scales, bits of hair, and other debris. Eggs hatch in 9 to 12 days, and an entire life cycle from egg to louse takes just 1 month.

The most common treatment for lice is over-the-counter residual pesticides designed for livestock. Louse control is difficult because pesticides kill lice but not their eggs. Because eggs hatch 8 to 12 days after pesticide application, retreatment is necessary 2 or 3 weeks after the first treatment.

Mites

Mange mites feed on the skin's surface or burrow into it, making teensy, winding tunnels from 1⁄10 inch to 1 inch long. They deposit a liquid at the mouth of each tunnel, which dries and forms scabs. Mites also secrete a toxin that causes fierce itching. Goats infested with mites sometimes rub and scratch themselves raw. Infestations are highly contagious; if one goat has mites, plan to treat your other goats, too.

Over-the-counter powders, creams, sprays, and dips are usually effective for treating mites. If you worm an infected goat with ivermectin, the ivermectin will also help control a number of external parasites, including mites. However, you need to know which type of mites you're dealing with so that you can make an informed decision regarding mite-control products. If you're unsure, take a skin scraping to your vet for a diagnosis.

Ticks

There are two kinds of ticks: hard ticks (Ixodidae family) and soft ticks (Argasidae family). Neither kind jump or fly, and their bites cause little, if any, initial discomfort. Both sexes attach and suck blood, but only the females engorge to many times their original size.

If you find a tick attached to one of your goats, take it off. The longer a tick is attached, the more likely it is that it will transmit any disease it's carrying. Ticks carry several potentially deadly diseases known to infect goats, including Lyme disease, ehrlichiosis, anaplasmosis, and Rocky Mountain spotted fever. Heavy tick infestations can also lead to serious anemia.

To remove a tick, use your gloved fingers, tweezers, a hemostat, or a commercial tick remover to grasp the tick's mouthparts as close to the goat's skin as possible and then pull straight back, slowly and steadily. Be careful not to squeeze the tick, because squeezing can inject whatever toxins the tick is carrying into the goat's blood. Kill the tick by dropping it into a container of alcohol or soapy water.

Many tick species mate while the female is feeding. After removing an engorged female, look closely for tinier males attached in the same location.

The best way to control ticks is by altering their favorite habitat: tall grass and weeds. Mow your pastures and keep your barn and outbuilding areas free of weeds and litter. Also consider keeping a flock of guinea fowl because ticks are a guinea's favorite meal. Free-range chickens are tick-eaters, too.

Flies

Flies, especially biting pests like stable flies, horse flies, hornflies, black flies, and midges, can drive goats (and goat keepers) to distraction.

Stable Flies

A stable fly looks a lot like a common housefly, but the stable fly has a stiletto-like proboscis that extends beyond its head and is used to pierce the host's skin and feed on blood. Stable flies prefer to feed in early morning and again in the late afternoon,

A stable fly uses its long proboscis to pierce its host's skin.

HOMEMADE FLY REPELLENT

You'll probably need to use fly repellent if biting bugs target your doe when she's on the milking stand. Commercial insect repellents formulated for horses or dogs, especially those made with natural ingredients, work reasonably well for goats. If you use one, read the directions carefully and avoid spraying or wiping it on bare skin, especially your goat's udder.

You can make your own fly repellent that will do a decent job without exposing your goats to toxic ingredients that could harm them. If you use one of these recipes, remember that essential oils can irritate delicate skin if not diluted properly, so don't add more than the recipe calls for.

- Combine 2 cups white vinegar, 1 cup Avon Skin So Soft bath oil, 1 cup water, and 1 tablespoon eucalyptus or citronella oil in a spray bottle. Shake before using.
- Mix 1 cup blue Dawn dish soap with 1 cup vinegar and 1 cup water. Shake well before using. The soap residue builds up over about a week, so wet your goat and bathe it off.
- Mix 1 part crushed garlic into 5 parts water. Place in a covered jar, shake, and leave overnight. Strain into a sprayer and use.
- Combine 3 parts Listerine and 1 part baby oil in a spray bottle; shake and use.
- Mix 3 cups apple cider vinegar and 1 cup water in a sprayer. Add 2 tablespoons lemon juice and 6 drops of tea tree oil and then shake well. Shake again before each use.
- Combine 1 cup strong black tea, 1 cup water, 10 drops of citronella oil, and 10 drops of lavender oil in a sprayer and shake well. Shake again before each use.

so keeping your goats indoors during the day can help protect them from stable flies. They prefer to feed on the lower parts of goats, such as their bellies and legs. Their entire life cycle from egg to adult is completed in 3 to 6 weeks. Many residual and knock-down insecticides are effective against stable flies.

Horseflies

Horseflies are the B-52 bombers of the biting-fly world—if you've been nailed by a horsefly, you know it! Horseflies have stout bodies and are up to 1½ inches long. A female horsefly will slash her host's skin with her sharp mouthparts and then lap up the pooled blood.

Horseflies are daytime feeders. They're especially active on warm, sunny days, when they're attracted to moving objects, warmth, and carbon-dioxide emissions. They skip from host to host to complete a meal, so they sometimes spread disease. Permethrin-based insecticides offer short-term relief for livestock, but other chemical repellents seldom work.

Horn Flies

Horn flies are teensy, roughly half to three quarters the size of the common housefly. Both sexes have painfully effective, piercing-sucking mouthparts. Adults spend most of their lives on a specific host, congregating on its back and shoulders or on its underside. They like hot, sunny conditions and feed up to 40 times per day. Females lay several hundred eggs during their short life spans. The entire life cycle from egg to adult is completed in

A goat's ears are prime targets for biting midges.

2 to 4 weeks, and several generations can hatch in a single summer. Pesticide sprays and dusts are fairly effective against horn flies.

Black Flies

Black flies are smaller versions of horseflies. Both sexes feed on nectar, but the female also drinks blood. Her bite is painfully out of proportion to her size. She slashes and then sucks pooled blood, injecting an anticoagulant that causes mild to severe allergic reactions. The swelling and itch that follow can last for 2 weeks or more. Hordes of black flies pose a serious threat to livestock, including goats. Mega-bitten hosts can die from anaphylactic shock.

Most black flies are daytime feeders that target animals' ears. Not much repels them; even DEET-based repellents are only minimally effective.

Midges

Midges, also called sand flies, punkies, and no-see-ums, are one of the world's tiniest biting flies. Only females suck blood. Bites are painless at first, but within 8 to 12 hours, tissues swell and an intense itch sets in. Midges are highly attracted to livestock, including goats, particularly to their ears and lower legs.

Most midges feed at dawn and twilight from early spring through midsummer. A few species are daytime feeders, especially on damp, cloudy days. Both types frequent salt marshes, sandy barrens, riverbanks, and lakes. Chemicals don't repel them, so to save your goats grief, keep them indoors during prime midge feeding time.

GOAT KEEPER'S NOTEBOOK

The Cat's Meow

According to Junwei J. Zhu, PhD, a research entomologist with the USDA's Agricultural Research Service (ARS), and Christopher Dunlap, PhD, a research chemist with ARS, catnip has been shown to repel more than 13 families of insects, including mosquitoes and stable flies. To try it, dilute 1 drop of catnip oil (buy it from aromatherapy retailers) with 15 drops of carrier oil and dab it on the hair-covered surfaces of your goat's body; you can also make a spray by mixing ¼ to ½ teaspoon of essential oil with 1 cup isopropyl alcohol and 1 cup water. Don't get either mixture in your goat's eyes or on her bare skin.

Horse Wormers

Livestock wormers tend to come in huge bottles and packages that go out of date before the average goat owner can use them up. However, you can buy paste wormers for horses that contain the same chemicals as goat wormers.

Quest horse wormer is made using the same chemical (moxidectin) as Cydectin. The standard dosage is 1 cc of Quest horse wormer per 100 pounds of goat. Several companies manufacture ivermectin paste wormers for horses, among them low-priced generics. The standard goat dosage is 1 cc per 35 pounds.

The easiest way to dispense a mini-goat-sized, teensy, measured amount from a horse wormer syringe is to squeeze the paste onto your latex-gloved fingers and wipe it on the back of your goat's tongue, taking care not to snag your fingers on his sharp back teeth.

CHAPTER 8

Health and Hoof Care

It's easier for you and better for your goats to keep them in the pink than to treat illnesses and accidents after they happen. And it's not that hard to do.

Taking Precautions

Start a routine of checking on your goats at least twice a day, even if they're out on pasture. Give each goat a quick once-over and make certain that they all appear to be healthy. Address sickness or injuries right away; don't wait to see whether your goat gets better by himself. If you don't know what's wrong with a sick goat, or you're not positive that you know how to treat him, get help.

At least once a week, walk around your goats' housing, exercise yard, and pasture, looking for and removing anything that could be dangerous to your goats, such as toxic plants, hornets' nests, stray hay strings, and the like. If you find sharp edges on metal buildings, broken fences, or any other sources of potential injury, make repairs right away.

Here are some more tips to keep your goats healthy and comfortable:

- Feed mold- and dust-free hay and grain.
- Install locks on feed room doors and double check to make sure that the doors are locked when you leave. Be aware that some goats are caprine Houdinis, and, if you have a goat like this, install multiple locks.
- Keep goat minerals and plenty of clean water available 24/7.
- Quarantine incoming and sick goats—no exceptions!
- Put an injured goat where his herdmates won't harass him, preferably in the company of a friend.
- Be there with a doe when she kids. Assemble a kidding kit (see Chapter 9) and know how to use it.

- Discuss vaccination and worming protocols with your veterinarian.
- Worm as needed and mark your calendar to keep vaccinations up to date.
- Buy from disease-free herds or have incoming goats tested for caprine arthritis encephalitis (CAE), caseous lymphadenitis (CL), and Johne's disease before they leave your quarantine quarters.
- Protect your goats from predators by building good fence.
- Never tether your goats unless you're right there to protect them.

Vital Signs

Teach yourself how to check your goat's temperature, pulse, respiration, and ruminal function. Learn how to measure these vital signs on healthy goats to avoid stress and confusion when you think you have a sick goat. Write each goat's normal statistics someplace where you can find them in a hurry so that you are able to compare them to the measurements you take when a goat is sick.

Kids' values are generally higher than those of adults, and external conditions can affect readings. Extreme heat and humidity, fear, and anger all elevate pulse and respiration rates, and a goat's temperature can be up to a full degree higher on hot, humid days. Plus, a goat's temperature rises slightly as the day wears on.

Temperature

A normal temperature range for a goat is 101.5–103.5 degrees Fahrenheit. To take your goat's temperature, you'll need a rectal thermometer. Veterinary thermometers are better because they're usually sturdier and faster, but a digital rectal thermometer designed for humans works, too. Traditional veterinary thermometers are made of glass and have to be shaken down firmly after every use to force the mercury back down into the bulb. A digital thermometer is faster, it beeps when it's done, and it needn't be shaken down—simply press a button, and it's reset.

When taking your goat's temperature, have a restraint or a helper on hand to keep the goat still.

Take your goat's temperature in three easy steps:

- Restrain him. It's easy to place a small kid facedown across your lap, but you'll have to either secure an adult to something sturdy, using a halter or a collar and a lead rope, or fasten him into a grooming or milking stand. Another option is to recruit a helper to hold on to your adult goat instead of tying him up.
- Use KY Jelly, Vaseline, mineral oil, or even saliva to lubricate the

CHOOSING A VETERINARIAN

Before you buy your goats, ask goat keepers in your locale which vets they recommend. When you have some names, visit their practices and ask the following questions:

- How much experience do you have with goats?
- Do you make routine farm calls? Will you come to my farm in an emergency? If not, do you have facilities to board my goat? What about after-hours and weekend or holiday emergencies?
- Will you mind if I research a problem in books or online and bring you my findings? (Some vets appreciate input; others take offense.)
- Will you mind if I do my own routine healthcare procedures, such as treating minor injuries and giving my own shots?
- Will you dispense prescription drugs, such as painkillers and epinephrine, if I need them?
- Do you offer payment plans for major procedures or collaborate with a lender that does?

Once you've selected a vet, don't wait for an emergency to try him out. Schedule a routine farm visit and see how that goes. Are you comfortable with his attitude and his work? If so, you have a vet!

A FIRST-AID KIT FOR GOATS

It's important to assemble a well-stocked first-aid kit and to know how to use what it contains. Keep the items in a sturdy carrier, like a tool box or a plastic food-service bucket with a lid, so you can grab it and run when you need it. Following are some of the essentials to include, but ask your veterinarian what else he suggests keeping on hand.

- Thermometer
- Alcohol wipes
- Gauze sponges
- Telfa pads
- 1-inch-, 2-inch-, and 4-inch-wide self-stick disposable bandages
- One or two 2½-inch wide rolls of gauze
- 1-inch- and 2-inch-wide rolls of adhesive tape
- Partial roll of duct tape
- Human-grade bloodstop pads
- Betadine scrub
- A 12-ounce bottle of saline solution
- Wound treatments (e.g., Blu-Kote, Schreiner's Herbal Solution)
- Probiotic paste
- Blunt-tipped bandage scissors
- Regular scissors, preferably with blunt tips
- Folding knife
- Hemostat or large tweezers
- Lantern-type flashlight
- Small LED flashlight

Keep in the house to use as needed:

- 3 cc and 6 cc disposable syringes
- ¾-inch and 1-inch 18- and 20-gauge disposable needles
- 1½-inch 18-gauge disposable needles
- 10 cc and 60 cc catheter-tip syringes
- Epinephrine (a prescription item)
- Additional pharmaceuticals as recommended by your vet

thermometer. Insert it about 1½ to 2 inches into the mini goat's rectum, depending on his size.

- Hold a glass thermometer in place for at least 2 minutes; hold a digital model in place until it beeps.

After recording the reading, shake down the mercury (glass models only), clean the thermometer with an alcohol wipe, and return it to its case.

Pulse

A miniature goat's normal pulse ranges from 60 to 90 beats per minute. The easiest way to check your goat's heart rate is with a stethoscope. Place it directly behind his elbow, listen, count the number of beats in 14 seconds, and then multiply by 4. If you don't have a stethoscope, you can take your goat's pulse by lightly pressing two fingers against the same spot behind his elbow or on the large artery about a third of the way down the inside of either back leg.

Respiration

A mini goat's normal respiration rate is in the range of 15 to 25 breaths per minute. Watch your goat's rib cage or flank and count the number of breaths he takes in 1 minute.

Place a stethoscope or two fingers behind the goat's elbow to take his pulse.

Ruminal Contractions

Ruminal contractions are the strong gut movements that send rumen contents on to the second compartment of a goat's stomach, his omasum. Conditions such as bloat and even simple indigestion cause a goat's rumen to slow drastically or stop working altogether, so it's important to know when this happens.

Check for ruminal contractions by placing your hand on your goat's left flank. You should feel one to four contractions per minute.

Easy Home Healthcare Procedures

Everyone who keeps goats should learn to give shots and administer medications. It may sound intimidating, but it shouldn't be. Ask your veterinarian or a goat-savvy friend to show you how, or follow these easy directions.

Give Your Goat a Shot

Nearly all goat shots are given subcutaneously (SQ; under the skin) and should be given using ¾-inch or 1-inch, 18 or 20 gauge needles. Some antibiotics must given intramuscularly (IM; into muscle mass). If your veterinarian tells you to give an injection intramuscularly, ask him to sell you the right syringes and needles and to show you how.

Use disposable plastic syringes. Disposable syringes are inexpensive, so use a new one every time, for every goat. Using a new needle each time is less painful for goats, and it eliminates the possibility of transmitting disease by way of contaminated needles. Don't try to boil and reuse needles because boiling damages the seal on the inside of the plunger.

Select the smallest syringe that will do the job. Small syringes are easier to handle than big, bulky syringes. You'll need a new needle for each goat, plus an 18-gauge, 1½-inch transfer needle to stick through the

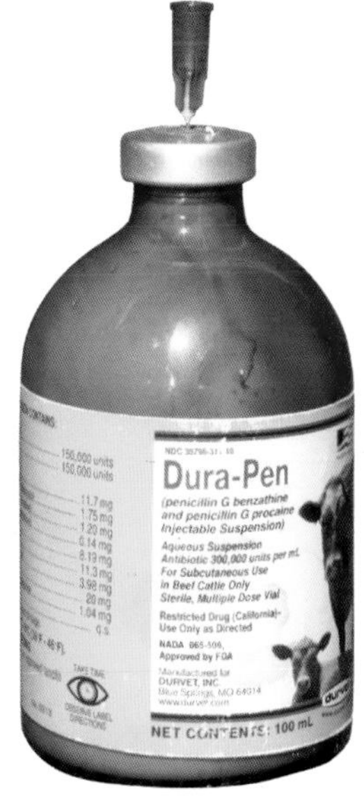

The transfer needle punctures the rubber cap on the medication bottle and is used to draw the medication into the syringe.

How Much Is a cc?

1 cubic centimeter (1 cc) = 1 milliliter (1 ml)
5 cubic centimeters (5 cc) = 1 teaspoon (1 tsp) = 1 gram (1 gm)
15 cubic centimeters (15 cc) = 1 tablespoon (1 tbsp) = ½ ounce (½ oz)
30 cubic centimeters (30 cc) = 2 tablespoons (2 tbsp) = 1 ounce (1 oz)

Break It Up

If you have to inject a relatively large volume of fluid, say more than 4 or 5 cc, break the dose into smaller amounts and inject them into multiple injection sites.

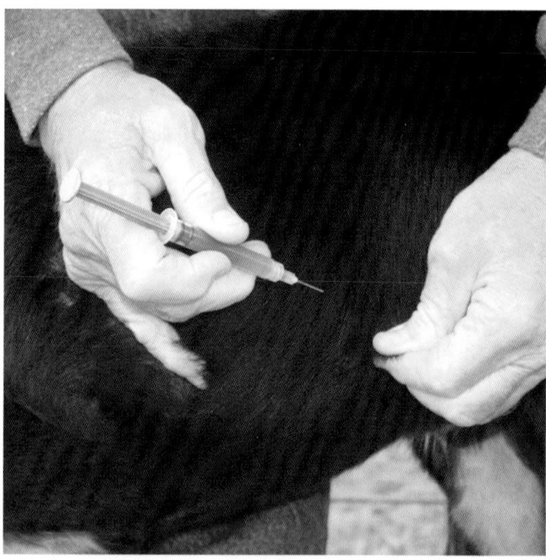

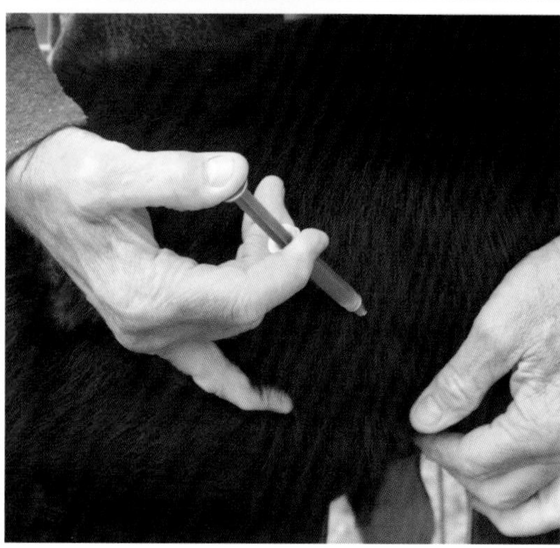

Pinch up a fold of skin (top), insert the needle into the skin fold (bottom), and depress the plunger slowly.

rubber cap on the pharmaceutical you're injecting and pull the vaccine/medicine out of the bottle.

Restrain your goat in the same way you would if you were taking his temperature. Jab a new sterile transfer needle through the cap of the pharmaceutical bottle; never poke a used needle through the rubber cap because this contaminates the whole bottle of fluid.

Attach the syringe to the transfer needle and then draw the vaccine or drug into it. For example, if you need to draw out 2 cc of fluid from the bottle, inject 3 cc of air to avoid having to draw fluid from a vacuum, and then pull a tiny bit more than 2 cc of fluid into the syringe. Next, detach the syringe from the transfer needle and attach the needle you'll use to inject the pharmaceutical into your goat. Press out the excess fluid to remove the bubbles created as you drew out the pharmaceutical.

Choose an injection site. Since injection-site nodules can be mistaken for CL abscesses, many people inject subcutaneous pharmaceuticals between the front legs, where CL abscesses don't normally occur. Other preferred sites are the loose skin on the neck, over the ribs, and in the armpit.

Never inject anything into damp or dirty skin. If your chosen injection spot is dirty, clean it with an alcohol swab.

To give the injection, position yourself so that you can work

EPINEPHRINE

Epinephrine is a prescription drug available through a vet. Also called adrenaline or "epi," its natural form is a hormone and neurotransmitter manufactured by the adrenal glands. Injectable epinephrine is used to counteract the effects of anaphylactic shock, a rapid allergic reaction which, if severe enough, can kill your goat.

Any time you give a goat a shot, no matter what product you use or the amount injected, you should be prepared to administer a dose of epinephrine to counteract an anaphylactic reaction, indicated by glassy eyes, slobbering, labored breathing, confusion, trembling, staggering, or collapsing. You might not even have time to fill a syringe. You have to be ready to inject the epi right away.

Many goatkeepers keep a dose of epinephrine drawn up in a syringe and stored in an airtight container in the refrigerator. It will keep as long as the expiration date on the epinephrine bottle, and it's easy to take with you whenever you give a shot. The standard dosage is 1 cc per 100 pounds. Be careful not to overdose epinephrine; it causes the heart to race.

comfortably. It's easier to bend over a mini goat than to squat at his side while you're giving the injection. If you're right-handed, stand on the left side and bend over his back to the right-side injection site. To give two injections, change sides.

To give an injection, pinch up a tent of skin and slide the needle into it, parallel to your goat's body. Try not to push the needle through the tented skin and out the opposite side or to prick the muscle mass below it. Slowly depress the plunger, withdraw the needle, and rub the injection site to help distribute the drug or vaccine.

You can use the same needle and syringe to inject the next vaccine or drug into the same goat, but you must use a new transfer needle for each new product.

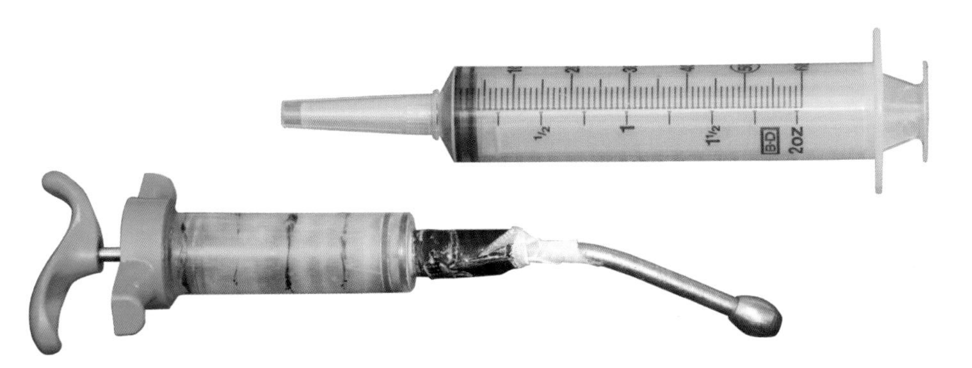

Catheter syringe (top) and dosing syringe (bottom). The nozzle on the dosing syringe has been reinforced with strong tape to prevent the goat from dislodging and swallowing it.

Hold the goat securely in place before inserting the syringe into his mouth.

Drenching Your Goat

You'll need to learn to dose your goat with paste and liquid medicines and wormers. This is called drenching. A drench can be given with a catheter tip syringe (not the kind you use to give shots), but the best way to drench is with a dose syringe.

To drench your goat, back him into a corner so he can't get away. Straddle his back, facing forward, and raise his head using one hand under his chin. With liquid substances, stick the nozzle of the syringe between his back teeth and his cheek; this way, he's less likely to draw fluid into his lungs than if you shoot it straight down his throat. Slowly depress the plunger, giving him time to swallow.

Place paste-type wormers or gelled medications on a gloved fingertip and deposit it as far straight back on the goat's tongue as you can reach. Keep his nose slightly elevated and stroke his throat until he swallows. And watch your fingers—goats' back teeth are sharp.

Give Your Goat a Pill

Ask your veterinarian if pills have to be given whole. If they don't, powder the pill using a mortar and pestle or by smashing it between two spoons. Dissolve the powder in liquid to give as a liquid drench or stir it into yogurt, jam, cold molasses, or chilled honey and give it as a paste. Note: Don't handle powdered pills with your bare hands because some drugs can be absorbed through human skin.

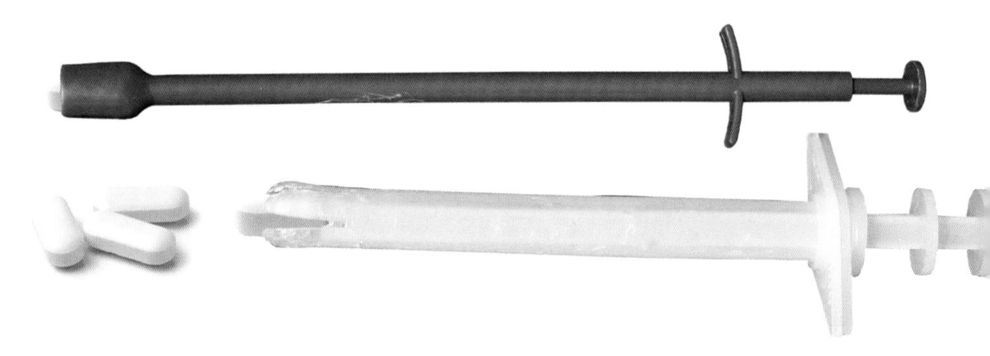

A balling gun (top) or a pet piller (bottom) is a handy tool for administering whole pills.

If the pill must be given whole, use a balling gun or a pet piller. Buy the smallest balling gun you can find or, if your goat is very small, a pet piller from the pet-supply store. A balling gun is a plastic device designed to propel a pill down your goat's throat. A pet piller is a scaled-down version of a balling gun designed to give pills to dogs and cats.

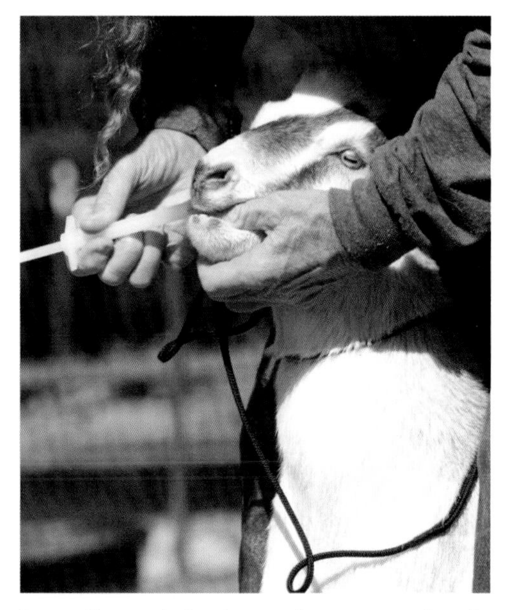

A pet piller made for dogs and cats can work well with a miniature goat.

Coat the pill with cold honey or molasses, jam, or yogurt and then place it in the balling gun and lay the assembly aside. Next, straddle your goat as though you were giving him a drench. Pry his mouth open, place the end of the balling gun or pet piller far enough into his mouth to deposit the pill at the base of his tongue, and raise his head to a 45-degree angle. Depress the plunger, pull out the balling gun, stroke your goat's throat until he swallows, and hope. If you're lucky, he'll have swallowed the pill. If he spit it out, pick up the pill, wipe it off, and try again.

Fourteen Diseases and Conditions of Concern to Goat Owners

Goats are prone to a host of diseases and afflictions that you'll probably never see, no matter how long you keep goats. Still, it's good to know about them, so you might want to buy a book like Cheryl K. Smith's *Goat Health Care: The Best of Ruminations* 2001–2007. It's an especially good reference, and it's written for owners of miniature goats. Another must-have veterinary-care book written for miniature goat keepers is *Pygmy Goats: Management and Veterinary Care* by Lorrie Boldrick, DVM. Following are some common goat health problems that you should know about.

Bloat

Bloat occurs when goats eat too much grain, too much legume hay when they aren't accustomed to eating it, or too much tender, high-moisture spring grass. Gas becomes trapped in a goat's rumen and expands until it presses so hard against her diaphragm that she'll suffocate without immediate treatment.

A bloated goat has taut, bulging sides. She breathes heavily, kicks at her abdomen, grunts, cries out in pain, and grinds her teeth. Bloat is often fatal. You must call your vet without delay when you notice the symptoms.

Any breed of miniature goat can run the risk of bloating.

In an emergency, if your veterinarian isn't available, a homemade bloat treatment that sometimes works is made of ½ cup of vegetable oil, ½ cup of water, and 2 tablespoons of baking soda, mixed well and given with a dosing syringe.

To prevent bloat, store grain and legume hay where your goats can't break in and overeat. And always feed your goats grass hay to take the edge off of their appetites before turning them out on lush spring pasture.

Caprine Arthritis Encephalitis (CAE)

CAE is an incurable viral infection caused by a retrovirus similar to the one that causes HIV in humans. Although there is a juvenile-onset form that causes seizures and paralysis in infected kids, CAE is primarily a wasting disease of adult goats.

Symptoms include swollen knees, lameness, progressive weight loss, a hard udder, and congested lungs. Goats with CAE eventually die of chronic, progressive pneumonia. Although not all CAE-positive goats develop symptoms, they can pass the virus along to their kids and to other goats.

There is no treatment and no vaccine to prevent CAE. However, because CAE is passed from does to their kids via colostrum and milk, breeders routinely remove kids from CAE-positive dams at birth and bottle-raise them using milk replacer or CAE-free pasteurized milk.

Caseous Lymphadenitis (CL)

Goats with CL develop thick-walled abscesses containing odorless, greenish-white, cheesy-textured pus. CL occurs on lymph nodes, especially on the neck, chest, and flanks, as well as internally on the spinal cord and in the lungs, liver, abdominal cavity, kidneys, spleen, and brain. CL is contagious and incurable. Transmission occurs via pus from ruptured abscesses.

Don't panic if your goat develops a lump. It could be an injection-site nodule or a pus-filled abscess caused by a puncture wound, splinter, or cut. The only way to know for sure whether an abscess is caused by CL is to have its contents cultured.

Any goat with a ripening abscess should be quarantined, and the abscess should be carefully drained and treated according to your veterinarian's instructions. Don't let drained pus contaminate your goats' surroundings; catch it in a latex glove and burn any pus not needed for a culture. CL is transmissible to humans, so it's important to wear protective clothing when lancing any abscesses. The goat should remain quarantined until her abscess site has healed.

There is currently no vaccine to prevent CL in goats, so it's best to buy from blood-tested, CL-free herds. CL-positive goats can be vaccinated with an autogenous vaccine custom-made from the pus of one of your infected goats, and although the goat won't develop new abscesses and can't spread the disease, she'll always test positive for CL.

"X" marks the spots where CL occurs on a goat's body.

Enterotoxemia

Enterotoxemia, also known as "entero" or overeating disease, is caused by bacteria found on most farms—in manure, soil, and even the rumens of healthy goats. Too much grain or milk, abrupt changes in quantity or type of feed, and drastic weather changes can all cause entero bacteria to quickly proliferate. When they do, they produce toxins that can kill a goat within hours.

Symptoms include bloating, standing in a rocking-horse stance, teeth-grinding, crying out in pain, seizures, foaming at the mouth, coma, and death. Treatment is usually ineffective because death occurs so quickly.

There are two types of enterotoxemia: type C, which is the type that quickly kills young kids, and type D, which affects slightly older kids and adult goats. The most common enterotoxemia vaccine is CD/T toxoid (the T is for tetanus), and it works for both types.

Goat Polio

Goat polio, also called polioencephalomalacia, isn't related to the viral disease called polio in humans. It's a neurological disease caused by a thiamine (vitamin B1) deficiency that causes brain swelling and, if left untreated, the death of brain tissue.

Symptoms include diarrhea, disorientation, depression, stargazing (the goat holds her chin up as though she's looking at the sky), staggering, circling, blindness, convulsions, and death in 1 to 3 days.

Thiamine deficiencies can be caused by moldy hay or grain, an overdose of the sulfa drug amprollium (CoRid) when treating for coccidiosis, certain toxic plants, and sudden changes in diet. Prescription thiamine injections from the vet can help a goat's condition improve in as little as 2 hours. As with all illnesses, though, prevention is better than cure.

Fusobacterium necophorum, *one of the components of hoof rot, is a normal presence in areas where goats are kept.*

Hoof Rot

Hoof rot, also called foot rot, is caused by the interaction of two specific bacteria. *Fusobacterium necophorum* is present in manure and soil wherever goats or sheep are kept. It's when the second bacteria shows up that hoof rot occurs. This second bacteria, *Dichelobacter nodosus*, can live in soil for only 2 or

3 weeks, but it can live in an infected hoof for many months. However, if you don't buy infected goats or sheep, you'll never have foot rot in your herd.

Symptoms include extreme lameness and the presence of a putrid, pasty goo between the outer surface of an infected goat's hoof and its softer inner tissues. To manage the problem, isolate infected goats, trim their hooves back to the affected areas, and remove as much of the rot as you can. Treat according to your veterinarian's instructions.

Hypocalcemia

Hypocalcemia, also called milk fever, is a metabolic disorder of does caused by a drop in blood calcium a few weeks before and right after kidding. It's sometimes confused with pregnancy toxemia. Both are life-threatening situations, so if you suspect either one, call your veterinarian right away.

Does with hypocalcemia lose interest in eating and become progressively weaker until they eventually lie down and won't get up. Treatments for hypocalcemia include drenching with energy boosters and dosing with oral or injectable calcium substances like calcium gluconate or CMPK (a common over-the-counter fluid calcium, magnesium, phosphorus, and potassium product).

To avoid hypocalcemia, provide your pregnant does with free-choice 2:1 calcium-to-phosphorous mineral mix and feed high-calcium alfalfa hay or pellets to late-gestation and milking does.

Johne's Disease

Johne's (pronounced YO-nees) is also called paratuberculosis. It's a contagious, progressively fatal, slow-developing bacterial disease of cattle, sheep, and goats, but it's most commonly seen among dairy cattle. It's caused by a close relative of the bacterium that causes tuberculosis in humans.

Johne's disease can enter your herd when an infected but healthy-looking goat is added to the group. It can spread to your other goats through oral contact with contaminated manure. Kids can also be infected by nursing infected dams.

Symptoms include diarrhea, progressive weight loss, and weakness, leading to death. There is no treatment and no vaccine to prevent Johne's disease.

Listeriosis

Listeriosis is caused by a common bacterium found in soil, plant litter, water, and even in healthy goats' guts. There are two types: one type causes paralysis and the other, more common, type causes encephalitis (inflammation of the brain).

Problems begin when dramatic changes in feed or weather conditions occur, causing bacteria in the gut to multiply. Parasitism and advanced pregnancy can trigger listeriosis, too.

Symptoms are much like those of goat polio and include disorientation, depression, stargazing, staggering, circling, one-sided facial paralysis, drooling, and rigidity of the neck in which the goat pulls her head back, toward her flank. If you suspect listeriosis, call your veterinarian right away.

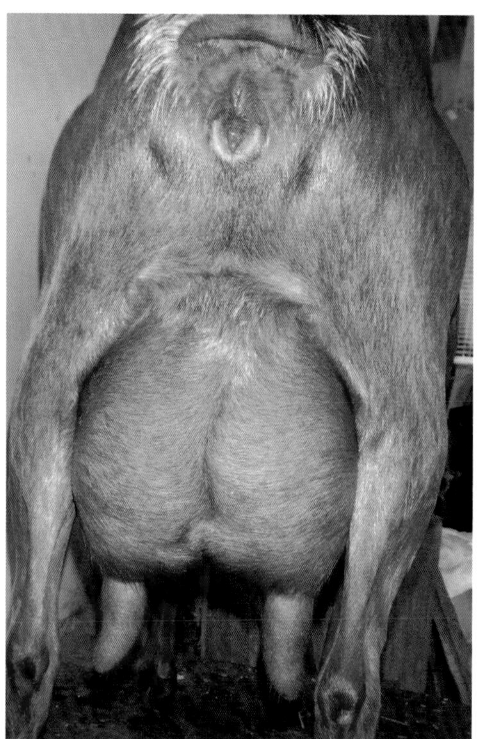

A healthy udder on a Mini Nubian, viewed from the rear.

Mastitis

Mastitis is a catch-all term for inflammation of the udder. It can be caused by a number of bacterial and staph agents, and it can also be triggered by substandard milking hygiene and delayed milking in dairy goats, udder injuries, stress, and milk buildup after does wean their kids. Untreated, it can progress to gangrenous mastitis and death. Symptoms include a hot, hard udder; teeth-grinding due to pain; lameness; loss of appetite; fever; decreased milk production; watery milk; and clumps, strings, or blood in the milk.

To help prevent mastitis, milk your does on a regular schedule and practice good milking sanitation. Reduce lactating does' grain rations for several successive days before weaning their kids. At the same time, switch them from legume to grass hay and eliminate all grain until their udders have dried up. If a doe has developed mastitis, the veterinarian will recommend suitable treatment based on the cause.

Pneumonia

Pneumonia is a serious and often fatal inflammation or infection of the lungs. It's caused by a variety of bacteria, fungi, and viruses that often gain a toehold due to environmental factors such as dusty feed, bedding, or surroundings; drafts or being hauled in open goat cages; stress; aspiration of milk, vomit, or drench into the lungs; and damage caused by lungworm infestations. Some forms of pneumonia are contagious, while others are not.

Symptoms include loss of appetite; depression; rapid or labored breathing; standing with forelegs out to the sides and neck extended; thick, yellow mucus; congestion; coughing; and an audible rattle in the chest. If you think your goat has pneumonia, call your vet.

Pregnancy Toxemia

Pregnancy toxemia, also called ketosis, is a metabolic disorder of does during their last few weeks of pregnancy and the first week or two after kidding. Old does, obese or thin does, and does carrying multiple kids are prone to pregnancy toxemia, but it can happen to any heavily pregnant doe.

A doe with pregnancy toxemia wants to be alone. She'll walk unsteadily, appear dull, and have little or no appetite. As the condition progresses, she'll grow increasingly weaker. She'll lie down and be unable to stand without help. She might have convulsions, grind her teeth, have trouble breathing, and go blind. Eventually, she'll die.

Pregnancy toxemia is caused by a disturbance in sugar metabolism but aggravated by stress. Because it is a potentially fatal condition, call your vet right away if you suspect toxemia.

Pregnancy toxemia can be prevented by maintaining does in good condition—neither too thin nor too fat—and feeding them a nutrient-rich, high-energy diet during the last 6 weeks of pregnancy.

Urinary Calculi (UC)

Urinary calculi are mineral salt crystals, also called urinary stones, that form in the urinary tract and can block a male goat's urethra, the tube that sends urine from his bladder to his penis. Does can also form stones, but they can pass easily through a doe's larger, straighter, and shorter urethra.

Symptoms include anxiety, pawing the ground, grinding teeth, crying out in pain, straining to urinate, rapidly twitching the tail, a swollen penis, mineral deposits on the hair around the penis, standing in a rocking-horse or hunched-over stance, and dribbling urine.

TREATS FOR URINARY-TRACT HEALTH

Unfortunately, ammonium chloride tastes extremely bad, so it can be hard to persuade goats to eat it. Miniature goat owner Michele Nelson shares these goat-tested and tasty treat recipes to help prevent urinary calculi.

Recipe #1

4 Tbsp. ammonium chloride
½ cup slippery elm bark, finely ground (available at health food stores)
½ cup peanut butter

Michele says, "Mix good and squish. I use my hands. I mold it into a square on the cutting board and cut it into twelve equal pieces. Roll each piece into a ball and put on a plate in the fridge."

Recipe #2

4 Tbsp. ammonium chloride
½ cup slippery elm bark
½ cup rolled oats
¼ cup molasses

"Mix and squish with hands and do the same as above. If the mix is too dry and crumbly to work with, I add a bit of water at a time until it's the right consistency," advises Michele. "Also, you can store these treats in a zip-top bag, but roll them in some extra slippery elm bark first so they don't stick to each other. Each wether gets one ball a day."

Urinary calculi require immediate medical attention. The condition won't correct itself, and, if left untreated, the goat's bladder will burst, and the goat will die.

Since castration stops both the production of testosterone and the growth of the urethra, many authorities recommend delaying castration as long as possible, giving the diameter of the urethra time to grow. It's also important to feed male goats a balanced 2:1 calcium-phosphorus ratio, which means feeding high-quality grass hay instead of alfalfa and offering little, if any, grain. Adding ammonium chloride to the diet may prevent most, but not all, types of urinary calculi from forming. Keep in mind that when goats don't drink enough water, urine becomes overly concentrated and crystals start to form, so male goats need access to clean drinking water at all times.

White Muscle Disease

White muscle disease, also called nutritional muscular dystrophy, is caused by a deficiency of the trace mineral selenium. Most of the land east of the Mississippi and much of the Pacific Northwest is selenium deficient. These are the areas where white muscle disease is most likely to occur, but other parts of the country are deficient, too. Your cooperative extension agent has maps that show mineral deficiencies in your area and is the person to ask about white muscle disease.

Hoof-care tools include thick gloves, a hoof plane, hoof trimmers (green handles), and hoof-rot shears (white handles).

The Perfect Hoof

Examine the hooves of a 2- to 3-week-old kid from the bottoms and from the sides. Trim at those angles and proportions; that's an ideal hoof.

Kids with white muscle disease are weak and may not be able to stand or suckle. Muscle tremors, stiff joints, and neurological problems are common, too. Symptoms in adults include stiffness, weakness, and lethargy. Even mildly selenium-deficient goats may be infertile or abort their kids, and they may have difficult births and retained placentas.

Injections of Bo-Se, a prescription supplement from your vet, often dramatically reverse symptoms, especially in newborn kids. All goats raised in selenium-deficient areas should be fed selenium-fortified feeds, have free access to selenium-added minerals, or be given BoSe shots under a vet's direction. To prevent kidding problems and protect unborn kids, does can be injected with Bo-Se 5 to 7 weeks prior to kidding. An important note: Make certain that you're living in a selenium-deficient area before treating goats for selenium deficiency because an excess of selenium is dangerous, too.

Hoof Trimming

It's not much fun for goats to have their hooves trimmed or for us to have to trim them, but it's an important part of keeping goats healthy and sound. Young goats' legs grow crooked to compensate for bad feet, old goats with untrimmed hooves are more likely to develop arthritis, and goats of any age that have to hobble around on overgrown hooves are susceptible to joint and tendon problems.

Trimming hooves is easy, but it takes a little know-how, too. You need the right tools to do a good job. There are more expensive tools, like the white-handled hoof-rot shears pictured on page 116, but the classic orange- or green-handled hoof trimmers designed for sheep and goats work best.

To get started, gather your tools and secure your goat. If you have a grooming or milking stand, place the goat on it. If you don't, use a collar and lead or a nonslip neck rope to tie him to a sturdy fence. Don't leave the lead so long that the goat can rush forward or yank back—about 10 to 12 inches of slack is perfect for mini goats. Crowd your goat against the fence to help secure him and then pick up a front hoof. If he tries to pull it away, be patient. It's scary and annoying to stand on three legs.

Use the tip of your hoof trimmers to clean dirt and muck out of the hoof. Some people use a horse hoof pick to do this job.

When the hoof is clean, begin trimming the outer hoof wall where it curls under the hoof and then snip any ragged bits you find between the claws of the hoof. Finally, nip off

any excess toe. If you trim hooves often, say every month or two, this is probably all you'll have to do. Occasionally, you'll need to pare the sole, which is the part between the walls on the bottom of a hoof. If you do, remove material in tiny snips and stop when you see pink; otherwise, you'll make the hoof bleed.

If you like, you can use a hoof or wood plane to clean up and further level the hoof. This step is optional, but it makes the hoof look nice. Goats' dewclaws also sometimes need trimming. If you trim them, make only tiny snips because the dewclaw will bleed if you take off too much.

Now go on around and trim the other hooves. Done!

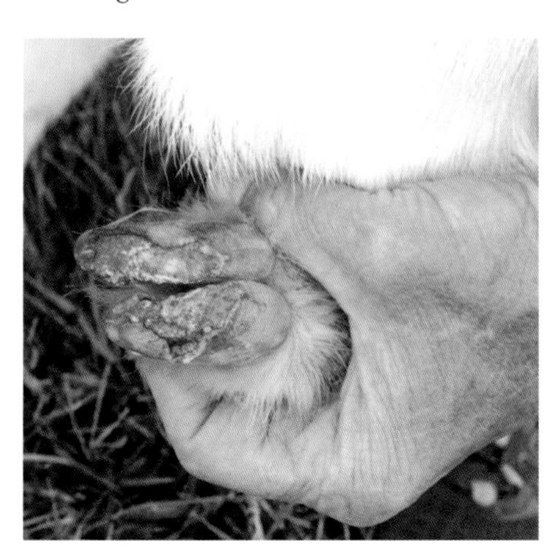

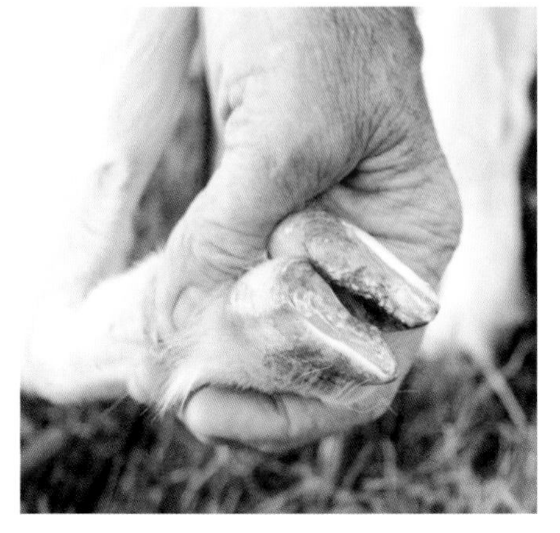

Left: a hoof that badly needs attention. Right: The same hoof after a thorough trimming.

GOAT KEEPER'S NOTEBOOK

Cleaning Wounds
Keep a supply of saline solution on hand for cleaning cuts, bites, and dings. If you run out of saline solution, flush wounds using lots of cool water from a hose. Apply a mild disinfectant, like dilute Betadine solution, to kill bacteria left on the wound. Be gentle. Don't scrub.

Wound Coatings
Most vets say that clean, open wounds heal better and faster than boo-boos coated with heavy ointments. If you like things natural, use holistic liquid dressings like tea tree oil or emu oil.

Honey Do
Unpasteurized honey is naturally antibacterial and makes a good dressing for cuts and scrapes. Saturate a piece of gauze with honey and fasten it over the wound. Even infected wounds respond quickly to this old shepherd's trick.

Hoof Soak
Instead of buying an expensive goat-sized soaking boot to treat an injured hoof, use a toddler's rubber boot from your favorite used-a-bit store. Shallow rubber ground feeders also work well, and most goats don't mind standing in them.

Make Your Own Ice Packs
Dampen the inside of newborn-sized disposable diapers and freeze them in individual plastic bags. To use one, wrap it around your goat's injured leg and hold it in place or secure it with disposable fabric wrap or an Ace bandage. You can also use frozen peas or corn inside a quart-sized freezer bag instead of a frozen diaper.

Another option is to combine 1 part rubbing alcohol and 3 parts water in a quart-sized zip-top freezer bag, filling the bag half full. Place a second zip-top bag over the first one to prevent leaks. When kept in a freezer, these reusable homemade ice bags stay icy cold but never freeze solid, and they conform nicely to a goat's leg.

Save Your Hand
If you're right-handed, always wear a thick leather glove on your left hand and vice versa. It's easy to slip and plunge the tip of your shears into your hoof-holding hand, especially if your goat jumps around.

Tough Toenails
Invest in a set of miniature horse hoof nippers if you have to trim really tough hooves. And because dry weather makes hooves really hard, plan to do your hoof trims a few hours to a few days after a rain.

Stay Sharp
With use, hoof trimmers eventually dull. Livestock-supply retailers that sell hoof trimmers usually also carry a handy tool to sharpen them. Instead of replacing trimmers, buy a sharpener and learn to use it. To help keep your hoof trimmers in good shape, clean them off and wipe a little bit of oil on the blades when you are done using them. Dirt and moisture left on the trimmers will eat away at the edges over time, which is why trimmers sometimes seem dull even after not being used.

CHAPTER 9

Breeding and Kidding

This chapter will tell you what you need to know about getting your miniature does pregnant and delivering their kids. In Chapter 10, we'll talk about caring for kids once they've arrived.

The Buck Stops Here

Your does will obviously have to visit a buck if they're to become pregnant. But should you keep a buck? It depends. Whether or not you choose to keep a buck, don't breed your doe to any old buck just so she can kid and come into milk. All kids are charming, but it's easier to sell them or to place them in good homes if they're quality kids with two quality parents.

Your Own Buck

One great advantage to keeping a buck is that you won't have to figure out if your does are in heat. If there's a buck on your property, they'll lustily let you know when they want to be bred. The other advantage is that you won't have to figure out how long your doe's heat cycle will be and then haul her to someone else's buck near heat's end, when she'll ovulate and be receptive to the buck. As a bonus, bucks are usually charming creatures. It's hard not to like a buck.

Still, the downsides to keeping a buck are many. You will, of course, have to feed him and provide for his other needs, but you'll also need to house him in secure surroundings away from your does but with another buck or a wether as his companion. He'll be incredibly aromatic during rut (breeding season, when hormones are raging!); some bucks even become amorous or aggressive toward humans when they're in rut. A 50-pound buck seems small until he's angrily smashing into you or leaping against you in a happy effort to help you make kids.

Weigh the pros and cons of keeping your own buck.

Buck behavior is often highly bizarre, especially during rut. A buck will strut around, flap his tongue, and vocalize by blubbering (it sounds as though he's saying "whup-whup-whup" or "what, what, what"). The musk glands on his forehead go into overdrive, producing stinky musk. And he'll shoot urine all over himself—on his face, belly, front legs, and even in his mouth—and also all over you if you stand too close. A male goat has a long, slim penis with a skinny, wormlike structure called the urethral process, colloquially referred to as the "pizzle," on its end. When a buck pees, his pizzle rotates like a pinwheel, sprinkling urine in all directions. If he wants to pee on you, which to him is a sign of affection, he will.

Someone Else's Buck

You can avoid keeping a buck by taking your does away to be bred to someone else's buck. If you choose this option, you must be certain that the destination farm is free of hoof rot and caseous lymphadenitis (CL) and that your doe will be cared for properly and housed separately from the farm's other goats if you have to leave her there.

MAKE A BUCK RAG

If you don't have a buck, and you aren't positive whether you'll be able to tell when your doe comes in heat, make a buck rag. Ask a friend who has a buck in rut to help you out by letting you visit his or her buck. When you do, take along a clean washcloth and a covered container big enough to hold the washcloth. Rub the washcloth all over the buck's scent glands—they're right behind his horns or where his horns would be—and the back of his front legs where he's peed on himself. Immediately pop the washcloth into the container and seal it up tight. When you suspect that your doe is in heat, whip out the washcloth and let her sniff. If she's in heat, she'll respond with tail-wagging excitement. Take her to the buck!

Urinating on himself is just one of a buck's bizarre mating behaviors.

Find out what the breeding fee covers. For example, will they charge you for breeding her through a single heat, or can you bring her back until she becomes pregnant? Does it cost extra if you have to board her with them for a few days until she comes in heat? If you do, can you bring your doe's usual feed if you want to? Work out the details before taking your goats to be bred.

How Do I Know if My Doe Is in Heat?

Heat is the period of time when a doe is sexually receptive to a buck. A doe in heat is restless, she bleats loudly and persistently, she urinates a lot, and she wags her tail rapidly (this is called flagging). She'll probably lose her appetite, her milk production could temporarily decrease, and the milk she produces may be oddly tangy. She'll probably rub against other goats—and you. Wethers and does higher in her herd's hierarchy may blubber like bucks and try to breed her. Her vulva will probably be puffier and redder than usual, and she might have a thin mucous discharge. If you have a buck, she'll want to hang out by his pen, and she'll object strongly when you take her away.

Goats and the Birds and Bees

Does come in heat every 18 to 24 days. Seasonal breeders, such as full-size LaManchas, Swiss dairy breeds, and most Nubians, only come in heat between roughly August and February. Aseasonal breeders, such as Nigerian Dwarfs and Pygmy Goats, come in heat year-round. Because most miniature goats are in part Nigerian Dwarf or Pygmy, they can be either seasonal or aseasonal breeders.

Heat lasts from 12 to 48 hours. Does ovulate near the end of their heat period, so this is when they should be bred. The two basic modes of breeding are pasture breeding, whereby the buck lives in the herd with his girlfriends, and pen breeding, in which both parties are taken to a pen, where they're released and the mating occurs.

Does in heat seek out and stay near the buck. Some sniff, lick, or nuzzle him as well. A buck approaching a potential girlfriend sniffs her urine, nudges her, flaps his tongue, blubbers, pants, and flehmens (curls his upper lip and grimaces in response to smelling the female nearby). If the doe is in heat, she'll encourage his advances while madly wagging her tail. If disinterested, she'll simply walk away.

Mating is achieved very quickly. When the buck ejaculates, he'll jerk his head back and slide off the doe. Watch closely or you'll miss it.

After the Fact

The most critical period for embryo survival is the first 21 to 50 days after breeding. In fact, embryos aren't implanted in a doe's uterus until about 50 days after she becomes pregnant, which is why it's important to avoid hauling, hoof trimming, showing, or anything else that might upset recently bred does. Also be sure to check vaccine and wormer labels; if a drug or chemical shouldn't be administered to pregnant does, use a wide felt-tip marker to prominently note that fact on each bottle.

A Pygmy doe can come into season throughout the year.

Casey, a Nigerian Dwarf doe.

Feed early- and mid-gestation does a maintenance diet; they shouldn't be thin or allowed to become obese. They don't need grain, but they do need good pasture or hay and access to goat-specific minerals.

Late-gestation does require additional energy, protein, calcium, selenium, and vitamin E. That means grain. Adjusting for her pre-pregnancy size, the actual or estimated number of kids she's carrying, and the quality and quantity of pasture or hay available, begin adding a small amount of grain to your doe's diet about 6 weeks prior to her due date, working up to roughly 3 cups (about a pound) per day, just before kidding.

A doe's calcium requirements more than double during late pregnancy. Grains are low in calcium, but some high-quality legume hays, especially alfalfa, can fulfill most does' calcium needs. It's best, however, to supply additional calcium through commercial grain mixes formulated for pregnant does (these also supply additional protein) or high-calcium mineral supplements. If you feed grain to your pregnant doe, be sure to feed legume (not grass) hay or add bagged alfalfa Chaffhaye or soaked alfalfa pellets to her diet to balance her calcium and phosphorus levels. This helps prevent hypocalcemia and pregnancy toxemia (see Chapter 8).

Additional selenium and vitamin E are also needed during late gestation. Low selenium levels are implicated in dystocia (difficult births) and retained placentas in does and white muscle disease in kids (see Chapter 8). You need to know if your pastures or the land on which your feed grows is selenium-deficient. If it is, feed high-selenium and vitamin E loose minerals year-round or give Bo-Se shots to your does. However, too much selenium is toxic, so it's important to discuss its use and dosage with your county extension service or veterinarian.

Your late-gestation doe also needs plenty of exercise. A good way to help her get it is to place feeders, water containers, and mineral tubs or dispensers in separate areas of your exercise yards, pastures, and paddocks so that she has to walk back and forth between them. Or walk her on a lead. She'll enjoy herself, and so will you.

Give her a CD/T toxoid booster between 5 and 7 weeks prior to kidding so that she will pass immunity to her kids via her antibody-rich colostrum (first milk). If you live in a selenium-deficient area, this is the time for her Bo-Se shot as well.

PACK A KIDDING KIT

Collect the things you need to help at kidding time and pack them in something you can sit on while the miracle of birth unfolds. A 5-gallon food-service bucket with a lid will work, but a cooler is a lot more comfortable. It should contain:

- A bottle of 7-percent iodine to dip kids' navels. This is considered a controlled substance that you'll have to get from your veterinarian, but it's worth the hassle. Alternatives include 1-percent iodine, Betadine, or Nolvasan (chlorhexidine).
- A shot glass to hold navel-dipping fluid.
- Sharp scissors for trimming long umbilical cords prior to dipping them in iodine. Keep these scissors disinfected and stored in a zip-top bag.
- Dental floss to tie off a bleeding umbilical cord if needed.
- A digital veterinary thermometer.
- A bulb syringe, the kind used to suck mucus out of human infants' nostrils.
- Disposable latex gloves.
- Two large bottles or tubes of obstetric lubricant, such as SuperLube from Premier1 Supplies.
- Betadine scrub for cleaning your doe prior to assisting her.
- A sharp pocket knife because you never know when you will need one.
- A hemostat.
- A collar and lead for the doe.
- A small LED flashlight that you can hold between your teeth or a clip-on LED flashlight for your hat.
- Clean towels for drying kids.
- Bottle-feeding and tube-feeding apparatus (see Chapter 10).

Ready, Steady, Go!

Stop milking your doe at least 2 months before she's due to kid. Milking takes a lot out of a doe, and her body needs time to rest up and prepare for kidding.

Begin watching for signs of impending birth as the big day approaches. A normal gestation runs from 145 to 155 days. Around 150 days is the norm for full-sized goats, but miniature does sometime kid slightly sooner.

The doe's udder will begin filling anywhere from about a month before kidding to immediately prior to delivering her kids. It's less noticeable with first fresheners (first-time mothers) and more so with does who have kidded before.

A week or two before kidding, the hairless area around a doe's vulva swells, the area along her spine sinks, and the base of her tail rises. About 24 hours before kidding, her vulva becomes longer, flatter, and floppier, and she might stream a string of mucus from her vulva. About 12 hours before kidding, you can pinch her spine at the base of the tail and almost touch your thumb to your fingers on the other side.

First-Stage Labor

First-stage labor generally lasts 6 to 12 hours for doelings and older first-timers, and 4 to 8 hours for experienced does. Uterine contractions begin during first-stage labor. The doe will be very restless and will try to drift away from the herd to seek a nesting spot. When she finds one, she'll dig a depression and then lie down, get up, circle, dig some more, and repeat the cycle over and over. She'll gaze off into the distance and possibly search for her unborn kids while murmuring in a soft, low-pitched, fluttery "mama voice." She'll yawn and stretch. She may pant, taking short, shallow breaths. Periodically, she'll stop eating and possibly grind her teeth, indications that she's in pain.

The fluid-filled bubble appears.

Second-Stage Labor

Stage two is the actual birth of a kid or kids. It generally takes doelings an hour to as long as 3 hours to deliver their kids. An older, experienced doe delivering a single kid should finish birthing in an hour or less.

As labor progresses and hard contractions begin, the doe will spend more time lying on

If the bubble has not yet broken, the first baby will be inside it.

Once the shoulders pass through, the kid slides out.

her side, with her chin up and her head in the air, or she may extended her head out in front of her. A large, fluid-filled bubble or water bag will appear at her vulva and then break. The first kid is gradually forced along the vagina until the tips of his toes and his nose appear at the vulva, possibly inside the bubble if it hasn't yet broken. The doe continues pushing, and once the kid's shoulders pass the vulva, he slides right out. As soon as he's born, his mother starts cleaning him, licking his face and head. When the birth of another kid is imminent, the doe leaves the first kid to deliver the next, and so on until all of her kids have been born.

The mother starts licking him clean immediately.

She does the same for the second baby.

Third-Stage Labor

The third stage of labor occurs when the doe expels the placenta (afterbirth). This usually happens within 1 to 4 hours after delivering the last kid. The placenta is a shiny, lumpy-surfaced, red piece of tissue streaked with white veins. There may be one placenta for multiple kids or one for each kid—either is normal. Some does eat their placentas if given a chance. This presents a serious choking hazard, so placentas should be promptly removed and buried or burned.

Your Part in this Process

You need to be there when your doe kids. Although it's true that goats have been having kids for thousands of years without human intervention, a lot of those goats and their babies died. Nature isn't always kind at birthing time. This is where you come in.

Of course, goats sometimes surprise their keepers and plop out kids right after they've been checked and their humans have gone back to bed. But do your best. A baby monitor in the barn is a good investment. Otherwise, if your doe is showing signs of first-stage labor, check her at least every 2 hours around the clock until she kids. Or bunk in her stall. Most goats like the company, and all you have to do to check her is open your eyes, roll over, and go back to sleep.

Psych yourself up so that you're ready to help deliver kids if you have to. Miniature does are no more prone to dystocia than big goats are, but because they are smaller, repositioning malpositioned kids is a lot trickier. If you have huge hands, find a willing friend with small hands to be on call, or be ready to call your vet. If you're the designated delivery assistant, trim your fingernails in advance.

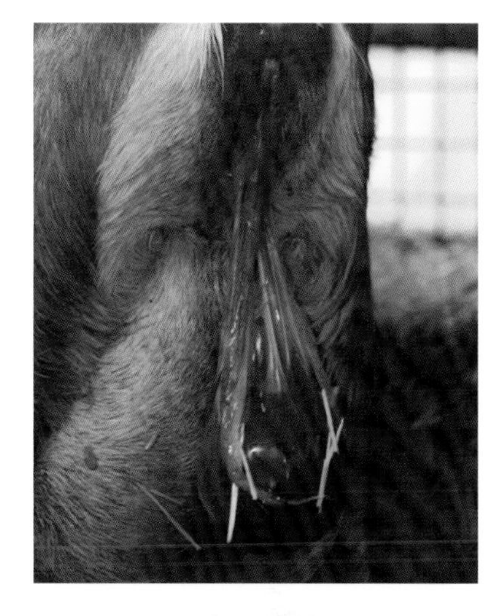

Top: Mom passes the placenta. Bottom: Mother and baby bond.

A Normal Delivery

Don't bother your doe if things go true to form; just sit back and watch the events unfold. When a kid is born, quietly ease over to it and strip the goo away from his face, especially his nostrils. If it's very cold out, lightly dry him off with towels from your birthing kit, but leave some goo on his body because cleaning it off helps a doe bond with her new baby.

Once the mama stands up, and the umbilical cord breaks, check to see how long it is. If it's

The Rule of Thirty

How long should you wait before assisting or calling your vet? Allow 30 minutes after hard labor begins for the birth membranes to appear, another 30 minutes for the first kid to be born, and 30 more minutes for second and subsequent kids to arrive. If things aren't resolved in that time frame, your doe needs help.

more than 2 inches long, use disinfected scissors to trim it back to 2 inches, and then dip the stub in iodine or another navel-dipping solution. Hold the kid upright, insert his cord into the fluid-filled shot glass, hold the shot glass tight to the kid's belly, and then quickly turn the kid over to give good coverage. Then get back out of the way and let the new family bond.

If the doe leaves her first kid to deliver another one, move the first baby out of the way but still close to his mother, where she can see him but not step on him. Repeat the process with each new kid.

When You Have to Help

Unfortunately, everything doesn't always go according to plan, so you need to be prepared just in case. Take some time to talk to your vet or read about kidding problems in detail so that you know what to expect.

Assisting with kidding sounds impossibly scary until you've done it a few times. You can correct basic problems yourself—yes, you can. However, keep in mind that the inside of a doe is very fragile, and kids cannot be forcibly pulled out the way that farmers pull out calves. You have to time your pulling to coordinate with the doe's straining and labor contractions. Once a kid is emerging, you have to always pull in a downward curve, toward the doe's udder, not straight back. Don't yank or pull suddenly and abruptly on the kid; instead, ease him from side to side or up and down to gradually get him out. Keep squirting lubricant into the vagina and around the kid if the birth canal becomes dry.

Know your limits and which presentations need to be handled by a veterinarian. If you need a vet, call him early—don't expect him to rush out and perform miracles after you and the doe are thoroughly exhausted. If a Caesarean section is necessary, delays only exhaust the doe and increase the risk of her vagina and uterus becoming dry, swollen, and bruised.

Once the Kids Are Born

Strong kids start seeking a teat within 10 to 30 minutes after birth. They'll poke around, looking for Mom's udder, and fall down a lot. This is normal. However, does sometimes have wax plugs in their teats that make suckling impossible, so take a squirt from each teat to make sure that both are clear and have fluid in them. Then make sure that each kid nurses. A kid will struggle if you try to place a teat in his mouth, so just watch and, if necessary, hold the kid near his mother's udder, where he can nose around and latch onto a teat by himself. If he's too weak to stand by himself or to suck, you'll probably have to take him into the house and care for him there. We'll talk about that in Chapter 10.

Growing up strong! Baby Dodger poses for a Christmas card.

After the last kid is born, what looks like a water balloon on a cord starts emerging from the doe's vulva, which means that she's beginning to deliver the placenta. By this point, she'll be tired and thirsty, so bring her a big bucket of tepid water with a dash of molasses or a small handful of sugar stirred in. Remove the bucket and any water that she hasn't slurped down, and your work is done. Check back every half hour or so until her placenta or placentas have been delivered; remove them before she eats them. Occasionally, you'll come out and find the balloon-like structure gone but no placenta in her stall, which means that she beat you to it and already ate the placenta. Not to worry. But try to be quicker next time.

Kids Need Colostrum

Colostrum is the thick, sticky, creamy-yellow "first milk" that a doe produces after kidding. In addition to being very nutritious, it contains high levels of antibodies against a variety of illnesses. A newborn kid doesn't carry antibodies because antibodies in his mother's bloodstream don't cross the placenta.

It's very important for kids to receive colostrum during the first 24 hours of life. Antibodies can cross the intestinal wall and enter the bloodstream of a kid only during this time period. Absorption is most efficient during the first few hours after birth.

Kids should ingest 10 percent of their body weight in colostrum within 24 hours after birth; this means that a 5-pound kid should consume 8 ounces of colostrum by 24 hours of age. Ideally, he should consume half of this within 4 to 8 hours of birth.

Does vary in the quantity and quality of colostrum they produce. Doelings produce less colostrum because they also produce less milk. Mature does have been exposed to more pathogens and usually have a higher concentration of antibodies in their colostrum.

All kids need colostrum. It's possible for kids to survive without it if they're kept in a relatively disease-free environment, like your home, but the likelihood of disease and death is higher in kids that didn't receive colostrum. Here are some things you can do if your kids are colostrum-deprived:

If a kid can't nurse from his dam but she's still available, milk her and tube- or bottle-feed her colostrum to the kid in 2- or 3-ounce portions until he's ingested enough.

If you can't use colostrum from the kid's dam, fresh or frozen colostrum from another doe will do. Sheep colostrum, especially from sheep kept on the same farm, is a very good alternative, too. Cow colostrum works, but since Johne's disease is widespread in dairy cattle, it should be used only if it's from a Johne's-free herd.

If no colostrum is available, the kid may survive if given a weight-appropriate dose of an immunoglobulin (IgG) replacer like Seramune Equine IgG, which large-animal vets are likely to have on hand. An IgG replacer is not the same thing as the inexpensive powdered supplements based on cow colostrum that you'll find on the shelf at your local feed store. If you have nothing else, you can try them, but they aren't very effective.

You can freeze colostrum for up to 1 year. Quick-freeze it in ice cube trays and, when frozen, pop the cubes into a double layer of freezer bags. Each cube is sized perfectly for mini kids' first meals. You can also freeze colostrum in 2-ounce portions in double-layered sandwich bags by popping the sandwich bags into a freezer bag for storage. Avoid storing colostrum in self-defrosting freezers. Their constant thaw and refreeze cycles affect its integrity. And never microwave colostrum. This kills the protective antibodies. Thaw frozen colostrum at room temperature or in a sealed bag or container immersed in warm water.

GOAT KEEPER'S NOTEBOOK

Bearding the Buck

If you find yourself in a sticky situation with an amorous or ornery buck, grab his beard and hang on tight. Walk him (he won't argue about it) to your nearest means of escape and don't let go until you're safe.

Buck Stink Be Gone

If you have to handle a stinky buck, wash your hands afterward with Gojo or Fast Orange hand cleaner from the auto parts store. They're the only readily available products that get buck stench off of human skin.

Let There Be Lights

Christmas tree lights strung in the rafters make wonderful lighting for kidding stalls. They throw enough light to make watching does and kids an easy task without being at all intrusive. And they're festive!

Pick a Pad

Keep a pack of heavy-duty mechanics' paper towels or a rolled stack of large puppy pads in your kidding box. Use them to dry off newborn kids. The puppy pads are also great for placing behind a doe as each kid pops out so that each baby lands on a clean, dry surface.

Goopy Laundry

The trick to getting goop and stains out of clothing worn while helping with kidding is to wash them right away instead putting them in the hamper for a future load. Smear blood and iodine stains with a big glob of laundry detergent, rub it in, let it stand for 20 minutes, and then wash the garment(s) in cold water. After a cold-water wash, immediately wash the clothes again, using hot water.

Fix that Ear

Kids, especially kids with pendulous ears, are sometimes born with a folded ear. Simply flattening the ears of a newborn kid and rolling them back and forth sometimes fixes this problem; do it every hour or so until the ear stays flat.

If it doesn't work in a day or so, make a brace by cutting two pieces of cardboard to the correct shape. If only one ear is folded, use the normal one to draw the pattern. Flatten the folded ear and sandwich it between the cardboard pieces, using adhesive or duct tape to keep them in place. If the ear sticks out, tape a coin to one of the pieces of cardboard for added weight before fastening it to the ear. Some kids don't like this and will work hard to get the braces off their ears, so check often and stick the brace back on if it comes off. It usually takes 3 or 4 days for the brace to work.

CHAPTER 10

Caring for Kids

Kids are arguably the cutest, sweetest babies in the animal world. Whether you buy a single bottle baby as a pet or raise the offspring of your mini milkers every year, miniature goat kids are so much fun to have around.

Let's assume that your mini milker has just given birth to healthy twins. Will you let her raise them (dam-raised kids), will you take them away and raise them yourself so you can have her milk (bottle kids), or will you go for a combination approach by allowing her to raise the kids but penning them separately at night so you can milk her in the morning? We'll talk about that last arrangement in Chapter 11.

Some say that dam-raised kids are wild and don't bond with humans the way that bottle babies do and that dairy-goat kids ruin their dams' sensitive udders when they nurse. Others say that bottle babies are clingy brats that can never be properly assimilated into a herd. None of these assumptions is true. If your doe is tame and you handle her kids from birth, they'll bond with you and will be tame, too. Udders, even the thin-skinned udders of modern dairy goats, are designed for nursing. And bottle babies who also interact with adult goats learn the ways of goats and humans.

Kid-Rearing Basics

Kid-proof the area where your kids will stay. Kids are bouncy and curious, and they can hide and climb better than you can possibly believe. Block crevices where they could get stuck, and make sure that they can't get wedged in gates or fences. Water your doe out of shallow water receptacles. Bowls designed for feeding large dogs work especially well. It's not unusual for a kid to hop into a deep bucket and drown.

Dodger's sweet face on his first day home with us.

Handle your kids to tame them. Kids, especially kids that haven't been handled a lot, are often frightened when you pick them up. Instead, sit on the ground and cuddle them in your lap. Scratch their chins; scratch their backs. Soon, they'll come running for attention. If you do pick a kid up, support him close to your body with your arm. Don't let him dangle.

By the same token, never roughhouse with male kids, especially bucklings destined for breeding or kids that will interact with humans as pet or therapy goats. When you bond with kids, you become one of their herd, and goats play rough. If you shove little bucklings around in play or you push on their heads and encourage them to push back, they grow up thinking that you like to play that way, too. What's cute when a goat weighs 3 pounds is not so sweet when he's a 60-pound buck. It's hard to correct play aggression in adult goats, so don't let the games get started.

Learn about kidhood illnesses so that if a kid becomes sick, you'll recognize the signs and will either know how to treat the problem or can take him to your vet. Consider coccidiosis (see Chapter 7). Without immediate, aggressive treatment, most kids with coccidiosis fade and die. To prevent this, most goat keepers treat their kids with a coccidiosis-prevention protocol. Ask your veterinarian.

Find someone in the know to disbud your miniature kids, and have it done as soon as their horn buds are big enough to disbud—most experienced disbudders say that's when you press the skin over the spots where a kid's horns would emerge and you can detect a bump that feels like a small pimple. Depending on your kid's breed and sex, that might be 3 days to a little over a week old.

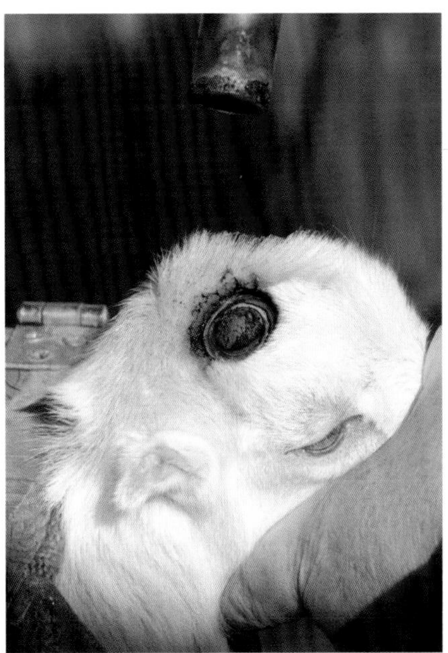

Disbudding is a delicate procedure that should not be attempted by a novice.

Don't try to disbud your kids based on something you've read in a book or seen on YouTube; always ask an experienced person to show you how and to guide you through your first disbudding session. Nigerian Dwarfs and breeds with a lot of Nigerian Dwarf in their background are notoriously difficult to disbud. If you burn too long, you will injure the kid's brain. If you don't burn long enough or do the job incorrectly, he will grow scurs—misshapen horn-like structures that aren't fully attached to the goat's head, so they often break off and bleed profusely. Small scurs are no big deal, but bigger scurs can be worse than horns. If they grow toward the goat's head, scurs have to be trimmed on an ongoing basis. Trimming large scurs is a difficult and bloody task.

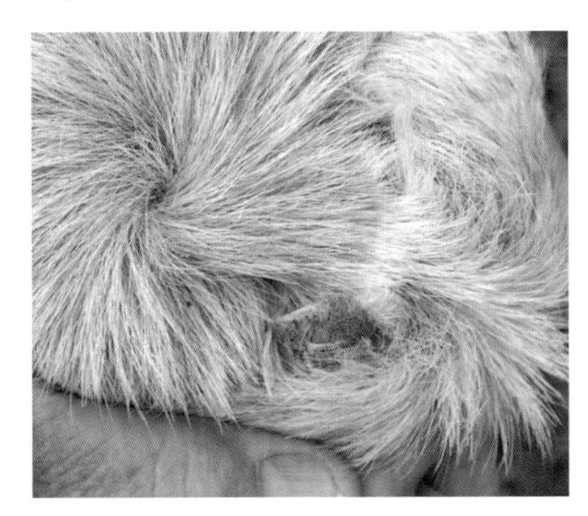

Simple (left) and serious (right) scurs.

Weaning time depends on the sex of the kid.

Dam-Raised Kids

If you dam-raise your kids, wean bucklings when they're 10 to 12 weeks old, or sooner if they're sexually precocious. Eight-week-old Nigerian Dwarf bucklings have been known to impregnate their mothers and other does in their mothers' herds. Doelings can be weaned at the same time if separated from their brothers, or they can be left with their mothers quite a bit longer—up to 4 or 5 months—if you feed your does well and don't need their milk. Some does wean their kids by themselves, but don't count on it.

Dam-raised kids start nibbling at grass and investigating their mother's hay and grain when they're just a few days old. Milk and what they eat at their mother's side is generally sufficient until they're weaned.

Bottle Kids

Bottle-raised kids are a different ballgame. When you take them on, you take on sole responsibility for their care.

If you're raising a group of bottle babies, say the offspring of your dairy does, you'll probably want to raise them in a barn or outbuilding. The exception would be if it's bitterly cold outside, at which point they'd need to be kept in your home, where it's warm, for a week or two. House groups of bottle babies indoors in large plastic tubs, playpens, or exercise pens designed for dog shows. A single bottle baby or a pair of kids will do well in the same type of housing, but large wire-type kennel crates, especially the kind that open from the top as well as from the front, are a better fit.

In either case, line indoor bottle-baby accommodations with pieces of blankets or large towels for bedding; tear or cut them into easy-to-launder sizes. They're warm and

ONE BOTTLE KID

Bottle kids are the exception to the "you must have at least two goats" rule. If you raise a kid as a pet and commit to providing for his emotional needs for the rest of his life, you can certainly keep just one goat because he will bond with you and your household pets instead of other goats. However, you can't lose interest and leave him alone and lonely for long periods of time. Goats raised by themselves eventually assimilate if rehomed and introduced to other goats, but that isn't the kind thing to do. If you raise an "only goat," expect to give him a loving home for life.

comfortable for the kids and easy for you to wash. If you find old wool blankets at yard sales or thrift shops, grab them! Wool soaks up a lot of urine and dries very quickly. Plus, the more you wash pieces of wool, the more they felt and the smaller and thicker the pieces get. Synthetic fleece is soft and wonderful for the top layer in a crate, and synthetic pressed-fiber blankets are good, too. Use several layers so that the bedding is nice and comfy.

Someone needs to watch kids the whole time they're loose in the house because they will get into everything they can reach. They'll chew telephone and computer wires, shred the toilet paper, and climb on the counters or your bed. But it's rewarding and entertaining. You'll see!

A miniature kid will adapt quite nicely to a warm, cozy crate indoors.

Dogs and Kids

Most pet dogs are good with kids as long as you're watching. Dogs are, however, wired to chase things that run. Running play can be misinterpreted by even the friendliest and best-trained dog, so don't leave your dog and a kid alone, unsupervised, inside or outside your home.

Housetraining Is Easier than You Think

You can easily housetrain a kid or two or three to go to the bathroom outdoors or on a puppy pad or in a litter box in your home. Why would you want to do that? Well, it's fun. And you'll save a lot of time washing and drying bedding, which you'll have to do constantly if you keep untrained babies in the house. Let's assume you want to train your kid or kids to the great outdoors. Here's what you need to do.

Housetraining a kid is exactly like housetraining a puppy, except the average kid learns much faster than most puppies do. The basic principle is that you keep the kid in his crate or playpen unless you're able to watch him closely. If he starts to pee in his bed or on the floor, scoop him up and carry him outside, set him down in the area where you want him to relieve himself, and make a big fuss when he finishes his business outdoors.

Always take your kid outside right after eating, napping, or racing around the house like a banshee. Carry him outside until he understands what you want; that way, he won't pee before you get him outdoors. Set him down in your chosen potty place and urge him to "go pee-pee" (or whatever your favorite phrase might be).

Housetraining a kid can be easy and fun. Ten-day-old Dodger shows how it's done.

The first few times you take him outside, it may take a while before he performs, but be patient. Don't interrupt him mid-pee, because he'll stop going. Just start with the happy "good boy!" chatter and, when he's finished, pick him up and cuddle him, play with him, or give him a small piece of animal cracker—something

DIAPERS ARE FOR KIDS

If you give your bottle babies free time loose in the house and they aren't housetrained, you'll probably want them to wear diapers. Doelings are easy: cotton or disposable diapers for comparably sized human infants do the trick, and the preemie size works for most mini kids. Be sure to snip out a hole for baby's tail—but don't make the hole too big. Instead of diapers, a buckling can wear a reusable, around-the-middle band designed for incontinent older male dogs.

Diapers tend to slip back as a kid frolics about. To prevent that, attach baby-sized suspenders to the diaper or pull a onesie on over the diaper. Both options work and are cute!

to show him that he's done well. If it's cold, rainy, or snowy, or for some other reason he's eager to go back in the house, take him inside as soon as he's finished. Otherwise, set him down and walk him around a bit in case he needs to "make berries."

Little kids can't hold urine very long, so you'll make a lot of trips outdoors at first. Fortunately, most kids "get it" pretty quickly and perform within minutes of going out. Once he understands, you won't have to carry him because he'll wait until he gets outside.

Bucklings are easier to train because most of them have a pre-pee ritual of backing up and then stretching out that tells you it's time to get him outside—quick! Doelings simply squat and pee, so you have to be quick to get them outdoors before they go.

With a young kid, take him out when you go to bed and again around 1:30 a.m., 3:30 a.m., and 6:00 a.m. By the time he's a few weeks old, one nighttime trip around midnight or 1:00 a.m. will be sufficient. Eventually, you'll have to take him out only at bedtime and first thing in the morning. If most housetrained kids absolutely must go in the middle of the night, they'll scream and wake you up.

For clicker training, you'll need a clicker to mark the behavior, a target, some tasty treats for rewards, and a handy carrier for the treats.

This housetraining routine works best if you clicker-train your baby. However, most newborn kids won't eat food rewards, so rely on lots of praise until they do. Some people click/reward using a slurp from a bottle of milk, but that distracts some kids—they're eyeballing the bottle instead of peeing.

Some kids poop only outdoors. Most kids poop mostly outdoors. A few want to poop wherever they are. Goat berries are easily picked up, even in the house, so don't obsess about that.

WHAT IS CLICKER TRAINING?

Clicker training is a gentle, reward-based system in which a trainer uses a hand-held clicker to tell her pupil that he's done well. The clicker signal is followed by a small but tasty treat. Equipment needed is minimal, and anyone with a basic understanding of how the system works can train any kind of animal to do just about anything.

All you need to get started is a basic understanding of clicker-training principles, a clicker (you can buy one online or at pet-supply store), a target, tiny pieces of your goat's favorite food, and an easily accessible reward carrier like a carpenter's apron or a small fanny pack. A target can be anything large enough to capture your goat's attention, like the marine float target in the picture, a dog's squeaky toy, or a plastic soda bottle.

Take your goat to a quiet place, away from obvious distractions. Click your clicker and hand him a tiny piece of food. Repeat this for a minute or two until he associates a click with a tasty reward.

Next, hold the target where your goat will accidentally bump it with his nose. The moment he does, click and reward. Don't demand perfection at first. If his nose touches anyplace on the target, reward him. Sessions should be short and sweet, five minutes tops.

Once he understands, refine the process. Hold the target up so he has to raise his head to touch it and then move it down near the ground so he has to lower his head. Once he follows the target, walk with him, rewarding him for following and touching it with his nose.

You're teaching him that when he does something correctly, he'll hear a click and earn a reward. The click is a bridge that tells him precisely when he did the thing you're rewarding him for doing. This is the basis of clicker training. Using it, you can teach your dairy goat to stand quietly on the milking stand, to lead quietly or pick up her hooves without a fuss, and even to pull a wagon or do tricks. If you can think of an action and break it into tiny steps that you can click and reward, you can teach it to your goat using clicker training.

If a kid has an accident in his crate, take the blankets out, spray the floor of the crate with dilute bleach solution, wipe up the bleach, and re-bed the crate with clean bedding. When you wash the soiled bedding, add a small handful of baking soda to the washing machine water to zap odors. It's important that the bedding doesn't smell like urine, or the kid may think that it's OK to pee in his crate.

If you're training more than one youngster (newborn to around 3 weeks old) at a time, try to scoop them all up and carry them outdoors at the same time, or else the ones left behind will put up a fuss and probably pee. As soon as they're old enough to accept food treats, lead the pack to the door and hand out a goodie to each one as they pass through. Soon, they'll all rush to the door when it's time to go out!

Puppy-pad and litter-box training are the same as outdoor training, except you'll take your kid to the pad or box instead. Once your kid understands what you want, he'll go to the proper pee-pee spot on his own when he has to go. It's that easy!

Feeding Bottle Kids

Milk, especially fresh or frozen goat's milk (see Chapter 11) is better for kids than milk replacers, which predispose kids to bloat and diarrhea. If you use milk replacer, choose a product made using real dairy whey instead of soy products, and make absolutely sure that it's specifically formulated for goats. Follow the mixing instructions exactly, and mix no more than one day's feedings at a time. If a kid has problems, try adding a bit more water than the instructions call for. Better yet, switch to real milk.

Bottle-Training Your Kid

Most newborns have a strong suckling reflex, but you'll still have to teach your kids to nurse from a bottle, which doesn't come naturally to them. Start by sitting cross-legged on the floor with the kid tucked between your legs, facing away from you, with his front legs straight and his butt on the floor. Raise his head by cupping your left hand under his jaw. Gently open his mouth. Insert the nipple with your right hand, steadying the bottle as you do; this way, he's less likely to spit out or otherwise lose the nipple.

Encourage the kid to lift and tip back his head. When his head is up and back, his esophageal flap closes. When his head is down, the flap stays open, and milk goes to the rumen, where it stagnates and can cause digestive problems. This is why

Three-week-old Nigerian Dwarf bucklings show off their strong suckling reflex.

THE NEONATAL DIGESTIVE TRACT

When a kid is born, his rumen is small, and the microbes present in an adult's rumen have not yet set up home. Instead, the abomasum is the largest of his four stomach compartments. The abomasum is the only compartment that contains digestive juices, so his digestion is more like that of a single-stomached animal than that of an adult ruminant. The closure of a flap of tissue in his throat called the esophageal groove channels milk directly to the abomasum instead of letting it enter the rumen and then the reticulum and omasum. When the kid starts to eat forage, his rumen, reticulum, and omasum gradually develop in size and function.

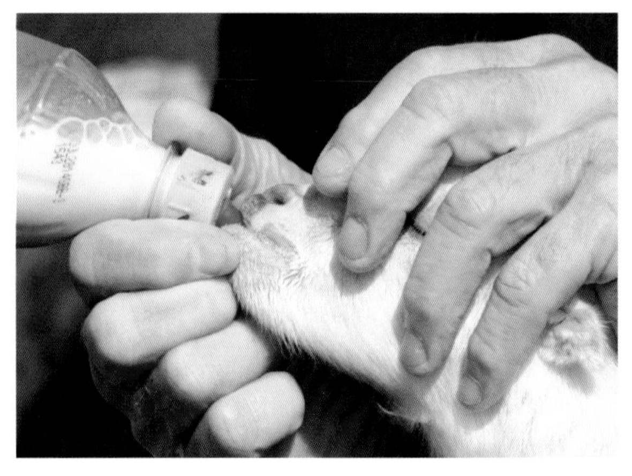

Bottle-feeding with a Pritchard teat and the kid's head tilted to the proper position.

it's better to feed kids from bottles than it is from troughs or pans set on the ground.

Don't let milk pour into the kid's mouth faster than he can swallow. Milk aspirated into his lungs can cause pneumonia. Hold the bottle as level as possible while still keeping milk in the cap and nipple.

If the kid you're bottle-training won't nurse, cup your hand over his eyes or place a towel over his head. This simulates the darkness of a doe's underbelly. If he still refuses, put him back in his crate or playpen and wait an hour or two before trying again. It's worth noting that kids that have previously nursed their mothers can be very hard to bottle-train. Persevere. If all else fails, tube-feed him. Washington State University's "Tube-Feeding Neonatal Small Ruminants" (see Resources) comes highly recommended.

Homemade and Healthy Milk Replacers

If you can't feed bottle kids real goat's milk, homemade replacers made with milk from the supermarket are better options than the commercial milk replacers. Here are two proven recipes you can use, but even plain homogenized full-fat cow's milk from the supermarket is better than commercial milk replacer.

Take It Easy

Abruptly switching from milk to replacer or vice versa upsets kids' delicate tummies. The result: diarrhea, also known as scours. Diarrhea leads to dehydration and serious problems, such as enterotoxemia, so if you switch, do it gradually over the course of several days. However, it's best to stick to the same milk or formula as long as you bottle-feed your kid.

Recipe #1

1 gallon whole cow's milk

1 12-ounce can evaporated (not sweetened condensed) milk

1 cup buttermilk

Pour about 1/3 of the milk out of the gallon jug and set it aside. Pour the evaporated milk and the buttermilk into the jug. Then fill the jug to the top using the whole milk that you set aside.

Recipe #2

1 gallon whole cow's milk

8 ounces dairy half-and half

Pour 8 ounces of milk out of the jug so you can replace it with the half-and-half. Shake. Done!

Feeding Supplies

Pritchard teats, like the one in the picture on page 144, are favorites with most people who raise miniature kids. The Pritchard teat's yellow plastic base can be securely screwed onto any 28-mm glass or plastic bottle. Plastic soda bottles and Pritchard teats make a good team because the bottles can be easily sanitized and reused several times and then recycled after a day or two.

Goat-supply companies, like Caprine Supply and Hoegger Supply Company (see Resources) carry Pritchard teats, as do many pet-supply and feed stores. Goat-supply retailers and pet-supply stores also carry preemie nipples and nursing kits that work well for tiny Pygmy and Nigerian Dwarf kids. In addition to nipples, buy a bottle brush, a measuring cup for mixing formula, and a funnel. Wash everything after every use, no exceptions.

How Much to Feed and When

Although goat keepers tend to strongly disagree about how often and what amounts to feed, this schedule is a good one to follow.

Days 1 and 2: 2–3 ounces of warm (102°F) colostrum per feeding. If you don't have real colostrum, use a milk replacer (commercial or homemade) with colostrum replacer added. Feed it six times at 4-hour intervals in 24 hours.

Days 3 and 4: 3–5 ounces of warm milk or milk replacer per feeding, six times in 24 hours. Also place a small, spill-resistant container or a crate cup of drinking water in the kid's quarters and change it frequently to keep it fresh.

Days 5 to 14: 4–6 ounces fed four times in 24 hours, gradually switching to milk or milk replacer fed either warm, at room temperature, or straight from the refrigerator.

Days 15 to 21: 6–8 ounces fed either warm, at room temperature, or straight from the refrigerator, four times in 24 hours. Introduce green pasture or free-choice leafy alfalfa or high-quality grass hay and small amounts of high-protein (16–20%) dry feed.

Days 22 to 35: Gradually work up to 16 ounces of milk or milk replacer fed three times in 24 hours. Continue until the kid is 6 weeks old.

Weeks 6 to 8: Make certain that the kid is eating sufficient amounts of forage (grass or hay) and chewing his cud. At 6 weeks, begin decreasing the amount of liquid offered at two feedings, eliminating them altogether by the end of week 8. Leave the remaining feeding at 16 ounces.

Weeks 8 to 12: Continue giving one 16-ounce feeding until the end of week 11; gradually eliminate the last feeding over week 12.

It's OK, though not necessary, to continue bottle feeding beyond 12 weeks of age. Some breeders feed milk once a day for up to 6 months. However, if your kid cries for food between meals, don't give him more milk. Overfeeding can quickly lead to enterotoxemia and bloat. Instead, give him a few ounces of easily digested Gatorade or a livestock electrolyte like Resorb.

Moving on Out

Eventually, you'll probably want to move bottle babies raised in the house out to your barn or goat shed. Be sure that the kids' outdoor quarters are warm, safe, well-ventilated, and free of drafts. If it's cold out, build "kid caves" using overturned cattle-sized mineral tubs or plastic totes with a door cut in one side of each, and bed them deeply with straw or clean hay that your adult goats have wasted. Several kids cuddled inside one of these caves generate a lot of body heat. You could, alternatively, fit each kid with a kid-sized dog sweater or goat coat. These are safer options than using heat lamps around flammable bedding.

A goat-sized sweater will help your little one stay warm outside.

Nipple Fatigue

Test nipples, especially soft rubber Pritchard teats, before every feeding. Tug on the rubber part to make sure it doesn't rip apart. When nipples fail, which they do after weeks of hard use, they usually fall apart all at once, and then your kid swallows the nipple. This doesn't always cause problems, but kids occasionally die when ingested nipples block their digestive tracts.

Sleepytime Pals

Place plush toys in your bottle baby's bed beginning on the day you bring him home. Most kids snuggle with soft toys when they sleep. A plush crib toy with a warm-in-the-microwave insert is an especially good choice. When plush toys get soiled or damp, put them in a pillowcase, tie the pillowcase shut, and pop them in the washer on the cold water setting; most can be laundered dozens of times.

Like a Parrot

Use a bed sheet to cover your bottle baby's crate or sleeping area at night. This cuts down on drafts in the wintertime and signals to the kid that it's time to go to sleep.

Ring Out the Bells

If the bottle kid you're trying to housetrain sleeps in a wire dog crate, suspend a string of too-large-to-swallow bells someplace inside the crate. He'll play with them, but, better yet, when he gets up from a nap, the crate will jiggle, and the tinkling bells will tell you that it's time to take him to his potty spot.

CHAPTER 11

Mmm-mm, Milk!

If you've never tasted fresh goat's milk, you're in for a treat. Once you've savored goat's milk, you'll never go back to tasteless cow's milk from the grocery store. And if you love goats, you'll find that home dairying is a fun, relaxing way to enjoy your animal friends. Is it worth your while? Yes!

Choosing a Dairy Goat

It's important to buy stock bred specifically for milk production. If someone in your area breeds dairy goats, ask if you can visit and pick his brains. *Dairy character*—the type of body build that indicates that a goat is or will be a good milker—is the same across the board, no matter what size or breed. Ask the experienced person to point out his best milkers and to show you production records (if he keeps them). He might even have a goat or two that you could buy.

Another good way to learn about dairy goat types is to join Facebook groups (see Resources) devoted to goat dairying. The Dairy Goat Conformation and Caprine Conformation groups are good learning venues, as is Miniature Dairy Goats, and there are groups devoted to specific breeds, including Nigerian Dwarf Goat Conformation, Nigerian Dwarf Goats—Dairy Side, Mini Alpine Dairy Goats, Mini LaMancha Breeders, and Miniature Nubian Dairy Goat Breeders. After joining, read posts and ask questions. Goat people are a friendly bunch and eager to interest others in their favorite breeds.

HOW MUCH MILK DOES A MINI GOAT GIVE?

The amount of milk a miniature goat is likely to give depends on many things, including her genetics (whether she's from high-producing bloodlines or pet stock), her size (a big F1-generation mini milker will usually outproduce a tiny goat of the same breed), what she's fed (grain-fed goats produce more milk than they would on pasture alone), how often she's milked (twice a day is standard, but low producers and does late in lactation are sometimes milked just once a day), where she is in her lactation cycle (she'll give the most milk about a month after kidding and taper off near the end of her lactation), and whether or not she's a first freshener (first fresheners, or first-time mothers, give less milk than they will during later lactations). All things considered, expect a well-fed miniature doe from milky bloodlines to produce in the neighborhood of 1½ quarts to 1¼ gallons of milk a day.

Once you've found a goat for sale, make an appointment to see her. Tell the seller in advance that you want to milk her so that he doesn't milk her out before you arrive. If you don't know how to milk, ask the seller to milk her while you watch. Does she stand quietly, or reasonably so, on the milking stand? Milking should be a relaxing interlude, not a battle.

Mystic Acres Blonde Ambition is a Mini Oberhasli with nice dairy character.

How fast does the milk stream out? It takes forever to milk a goat with tiny orifices. Grasp her teats. Mini breeders are selecting for larger, more user-friendly teats, but the fact remains that many minis have teats so small that they're hard to milk. Keep in mind that first fresheners have small teats, but they'll grow longer and thicker as she feeds kids or as she's milked. Other than that, what you see is what you get.

It doesn't happen often, but some does give strong-tasting, off-flavored milk. Ask for a taste. You want to make sure you like what she delivers.

About Goat's Milk

Milk and dairy products from healthy, well-fed goats are yummy treats. Properly handled goat's milk neither smells nor tastes "goaty." In fact, it tastes like whole-fat, home-processed cow's milk.

The nutritional differences between cow's milk and goat's milk are negligible. Goat's milk is slightly higher in calcium, milk solids, and a few vitamins and minerals, but the protein and carbohydrate counts are much the same. However, smaller fat globules make goat's milk easier to digest and whiter because it lacks the carotene that turns the fat in cow's milk a pale, creamy yellow (goats convert carotene to vitamin A).

Milking 101

Your goat will have to be bred and give birth before she begins producing milk. Most dairy goats are bred every year. Some, but not many, goats, especially goats of the Swiss dairy breeds, "milk through," meaning that they are capable of milking for a longer time before being rebred.

Goats must be milked every day at the same times, preferably in the same place and by the same milker. They never take long weekends or sleep in.

Plan to milk your goat twice a day, usually at 12-hour intervals. However, if this doesn't suit your lifestyle, you do have options. Instead of permanently separating your doe from her newborn offspring so you can milk twice a day, allow her to raise her babies and milk her just once in the morning.

To do this, pen her kids separately at night, preferably in a pen in her stall so they can still be together, and milk your goat first thing in the morning. After milking, let the kids out to be with their mom. They can then nurse until evening, when they're shuttled off to their separate quarters again.

NOW AS THEN

The following is from a favorite old book, *The Book of the Goat* (see Resources) by Stephen Holmes, published in 1909. When it comes to selecting good dairy goats, not much has changed since then.

"Always look for a milch [milk] goat with a long body and a large deep frame, the ribs being well rounded, so there is plenty of room for a big stomach. A heavy milker is generally wedge-shaped [when viewed from the side]—that is to say, it is much deeper in the hindquarters than at the chest.

"[The udder] should not only be large, but thin in substance and soft and elastic to the touch. When quite full it will be greatly distended, but after milking it should shrink up to a very much smaller size.

"The teats should be situated fairly well apart from each other and point down or slightly forward, the nicest being those that are long and tapering. The udder should by preference be round rather than long and narrow."

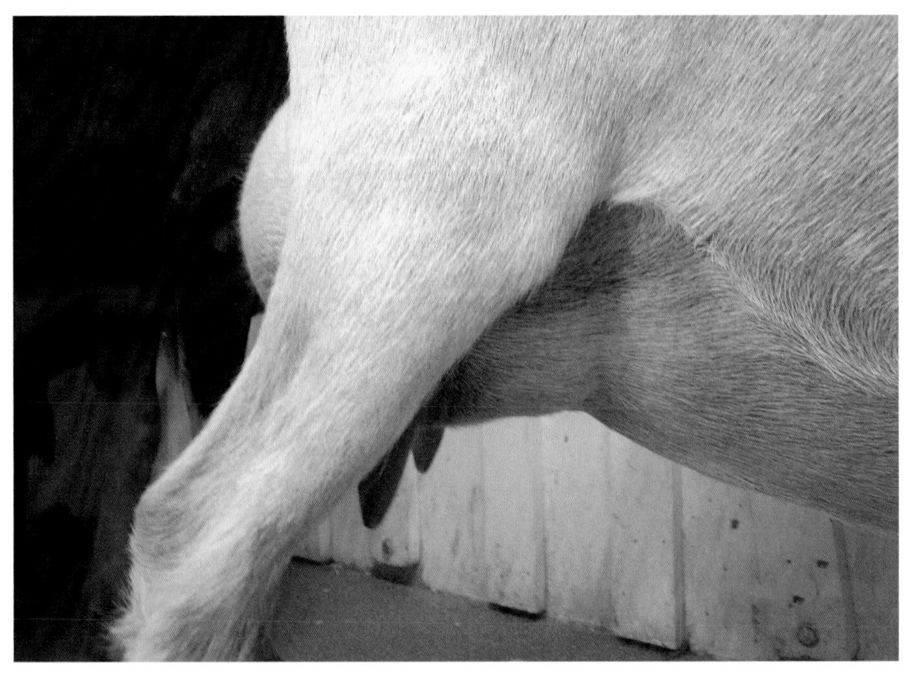

A side view of a nice udder on a Mini LaMancha.

Is this cruel? Not at all. Modern miniature dairy goats are bred to give a lot more milk than their kids require. Babies can be fed good hay and a little grain in their own pen, and, in many cases, they'll grow better than if they were raised 24/7 with their mother.

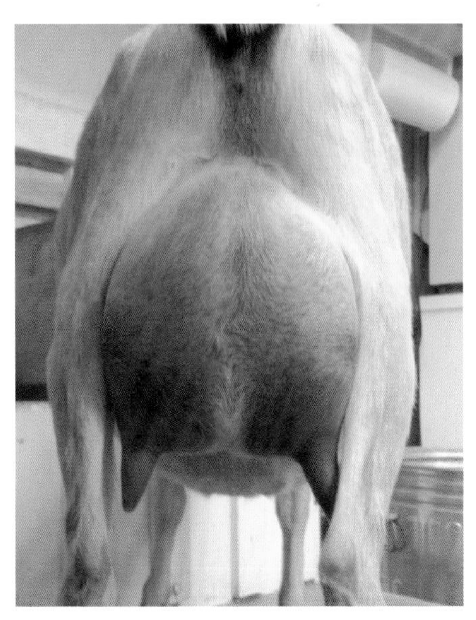

A rear view of nice udder on a Mini Nubian.

The amount of milk that your goat gives will decrease as her lactation progresses. She'll also need a period of downtime, usually 2 months, between lactations, when she won't be milking at all. To count on a steady supply of fresh milk, you'll need more than one milk goat.

You'll ideally need a separate milking area, although you can milk in your goat's stall if there is no other suitable place available and you stress cleanliness at every step. You'll also need the right equipment (we'll talk about that a bit later).

A goat has two teats attached to the halves of her udder, the udder being the entire mammary structure. Between milkings, milk accumulates in your goat's udder in structures called alveoli. They

UDDER CONSIDERATIONS

Buy a doe with a good, well-attached udder, meaning that it is held up high against her body and not flopping around, with a well-defined medial suspensory ligament down the center that visually divides it into two halves. An udder should be globular in shape with nicely shaped, decent-size teats placed in the middle of each udder half, pointing down toward the milk bucket.

A dairy doe should have two teats, each with a single orifice—no more, no less. Small extra teats (supernumerary teats) and nubbin-like projections are not uncommon but should be avoided, especially if a supernumerary teat has an orifice in it. If it has an orifice, it's probably functional, and if you don't milk it out—sometimes a difficult task due to its placement and size—it will be prone to developing mastitis. But the main reason to avoid them is because supernumerary teats are said to be hereditary, although their exact mode of inheritance isn't known.

then pass through a series of ducts into the gland cistern, the udder's largest collecting point. The gland cistern is connected to the teat cistern, a cavity inside each teat where milk pools until milking time. Circular sphincter muscles surround the orifice at the tip of each teat. When an external force, like a kid's mouth or a milker's hands, overcomes the strength of the sphincter muscles, they open, and the stored milk flows out.

When you prep your goat by washing her udder, the hypothalamus in her brain signals her posterior pituitary gland to release oxytocin into her bloodstream. This causes tiny muscles around those milk-holding alveoli to contract. In other words, she "lets down" her milk.

However, if your goat gets angry, hurt, or frightened, her adrenal gland will secrete adrenaline. This constricts the blood vessels and capillaries in her udder and blocks the flow of oxytocin needed for milk letdown.

Good hand-milkers are efficient and patient. They approach milking in a low-key manner, and they practice good milking technique. You can do it, too! Really, you can.

Udder Cream

You don't have to put udder cream on your goat after dipping her teats, but it's nice if you do. A goat's udder swells with milk and deflates after milking twice a day. Creams keep it pliable and less likely to chap or crack.

You will need:

- clean hands with short fingernails
- a sterilized, stainless-steel pail or bowl
- udder wash and paper towels or a box of unscented baby wipes
- teat dip or an aerosol product like Fight Bac
- a teat-dip cup or disposable 3-ounce paper cups (if you use teat dip)
- a strip cup with a dark, perforated insert or a dark-colored bowl
- a milking stand set up in your milking area

Lead your goat to the milking stand. Ask her to hop up, and then secure her head in the stanchion. If you feed her on the milking stand, this is the time to feed her. Sit down next to your goat on a milking stool or on the stand itself.

If you've never milked before, take a deep breath. When you're first starting out, milking can be scary, especially if your goat is a first freshener and doesn't know the ropes either. Goats pick up on their humans' vibes, and your doe can get frightened or impatient if you're upset. Stay calm; this will help her stay calm. In fact, sing! You can't nervously hold your breath if you're singing. And most goats seem to like to hear people singing.

Swab her udder and teats using udder wash or unscented baby wipes and then massage her udder for 20 or 30 seconds to facilitate milk letdown. Squirt the first few streams of milk from each teat into your strip cup or dark-colored bowl and examine it for strings, lumps, or a watery consistency that might indicate mastitis, and then dump it out. You aren't wasting milk when you do this. Most bacteria in milk is in the first few squirts, so you'd want to throw it away anyway.

Gather several clean receptacles, wipes, a teat disinfectant, and some treats before you get your goat on the milking stand.

A MINI-SIZED MILKING STAND

You can buy commercially made metal milking stands, but they aren't ideally sized for smaller miniature goats. If you want a stand to fit your minis, try a wooden stand made by handcrafters (for example, at Etsy; see Resources) or build your own using plans you can find online. Two especially good plans are listed in the Resources.

When choosing plans, keep your goat's comfort in mind—she shouldn't have to stretch to reach the stanchion and her feed cup—as well as your own. Don't build or buy a stand with a deck so wide that a nervous or ornery goat can tap dance sideways out of your reach.

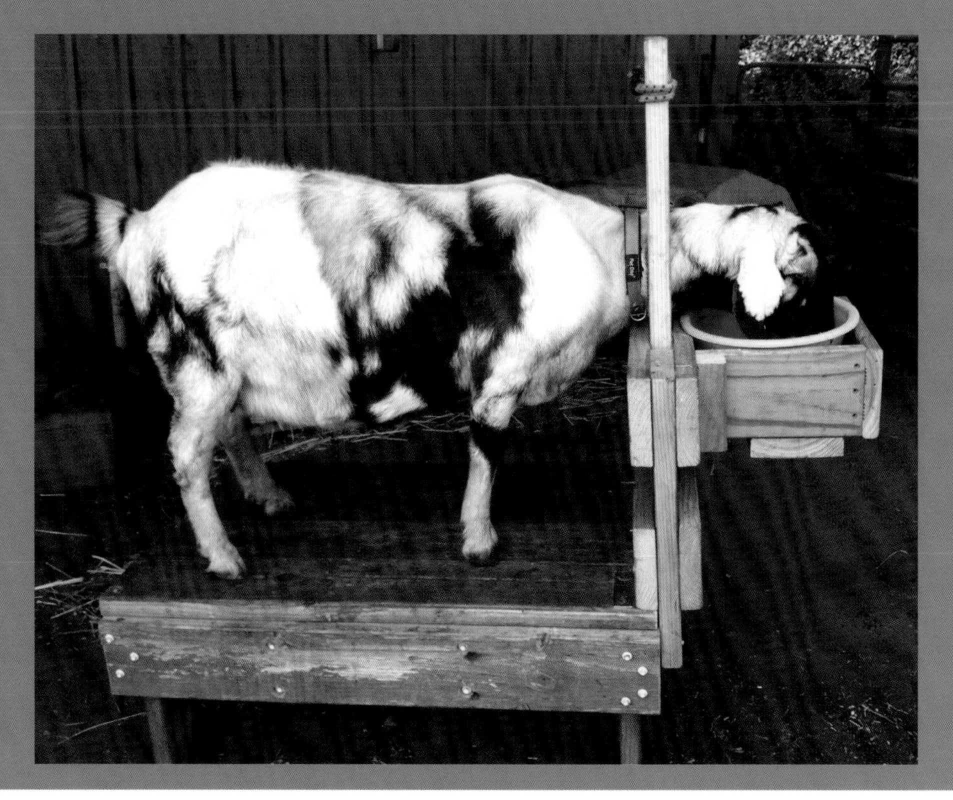

Place the milking pail slightly in front of your goat's udder and grasp a teat in each hand. Trap milk in each teat by wrapping your thumb and forefinger around its base. Gently nudge the doe's udder with the upper edge of the same hand and close off the teat with your thumb and forefinger. Never allow milk to backflow into the teat, and be sure to grasp just a teat and not the lower part of the udder itself, because pinching the udder on an ongoing basis can damage it. Plus, it hurts the goat.

Due to teat length in mini goats, you'll probably milk with two or three rather than four fingers. Still pinching with your thumb and forefinger, squeeze with your middle finger

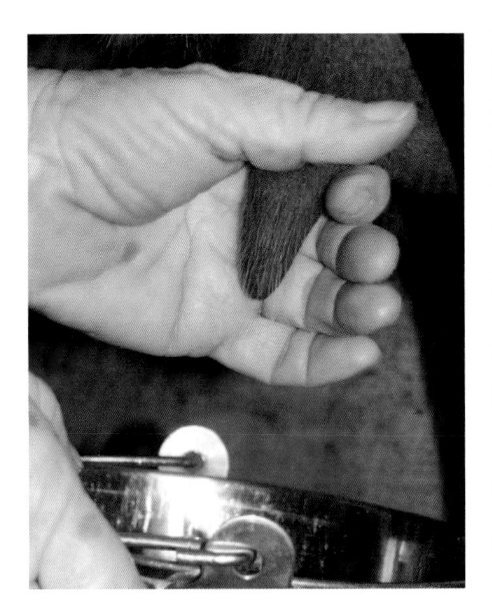

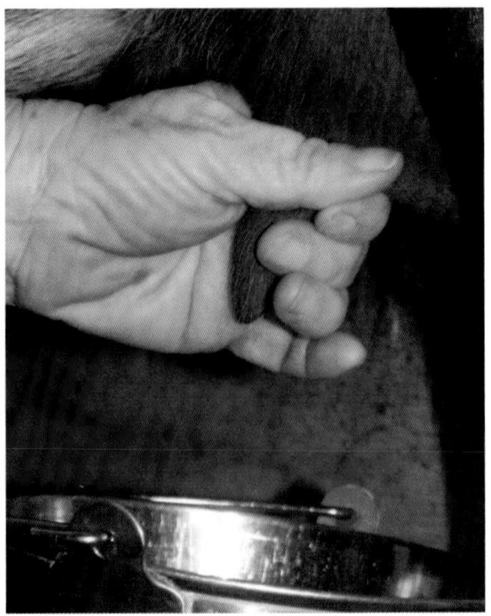

and then your ring finger (you'll be holding your pinky finger out to the side) in one smooth, progressive motion to force milk trapped in the teat cistern out into your pail. Relax your grip to allow the teat cistern to refill and, holding your hand in place rather than nudging the udder again, quickly repeat the motion: middle finger, ring finger, relax, pinch; middle finger, ring finger, relax, pinch. Congratulations, you're milking!

You can squeeze both teats at the same time or alternate by squeezing one teat while the other refills. If your goat is a kicker or a nervous first freshener who might knock over your pail, milk one-handed into a smaller stainless-steel container, dumping the milk into your milk pail as the container fills up.

Gently bump or massage the goat's udder to encourage additional milk letdown as the udder and teats deflate. Milk the goat until her udder is empty. If you consistently leave milk in her udder, it signals her body that the milk isn't needed, and her production will fall off accordingly.

Don't finish by stripping the teats between your thumb and fingers. This hurts and annoys the goat. By the same token, never pull on her teats. That hurts, too.

When you're finished milking, pour enough teat dip into your cup to dip one teat into fresh solution. Dump it, refill it, and then dip the other teat. Allow both teats to dry. As an alternative to dipping, spray the end of each teat with an aerosol spray like Fight Bac until a bead of fluid forms on each tip. Then release the goat and let her jump down from the stand.

Quick-Cooling Milk

Pour the milk through a stainless steel strainer lined with a milk filter into clean glass containers with lids. Quart and half-gallon canning jars with lids are perfect for storing milk in the refrigerator. You can buy milk filters at farm stores or from the goat and dairy

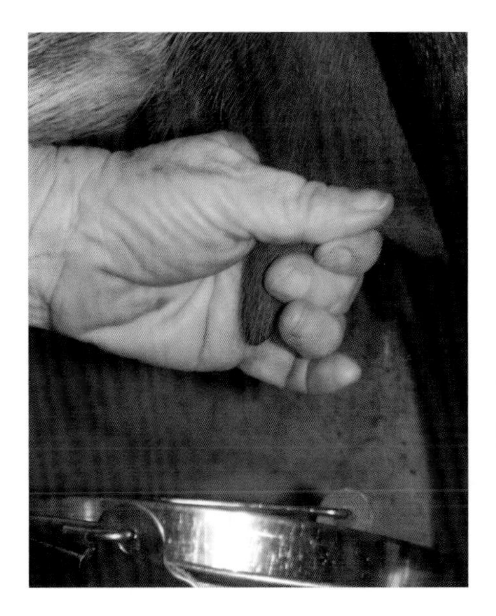

Milking sequence showing proper hand position.

supply retailers listed at the back of this book. You could use coffee filters in a pinch, but they cost more and don't work as well.

Cool the milk as quickly as you can. Some people put milk containers in the freezer for 10 minutes and then transfer them to the refrigerator for further cooling. Others immerse the containers in ice water in the sink—you can freeze water in secondhand plastic containers to make large ice cubes to cool the water faster. After 15 minutes or so, dry off the jars and either pop them in the refrigerator if you use your milk raw or pasteurize your milk before you store it. Raw milk should stay good in the refrigerator at 40°F or cooler for 5 to 7 days when handled this way. Pasteurized milk stays good for up to 2 weeks.

How to Pasteurize

If you decide to pasteurize, you'll probably want to buy a pasteurizer because pasteurizing on the stovetop is exacting work. If you must do it on the stovetop, though, here is a summary of two home-pasteurization methods from the South Dakota State University Cooperative Extension Service's bulletin "Home Pasteurization of Raw Milk" (see Resources). Using either method, you can pasteurize a quart to a gallon of milk at a time.

Method 1:

- Prepare an ice-water bath for cooling your milk.
- Place the raw milk in a double boiler.
- Heat the milk to 145°F and maintain that temperature for 30 minutes. If the temperature falls below 145° F, restart the process.
- Cool the milk as quickly as possible by either cooling it to 40°F within 4 hours in the ice-water bath or by cooling it to 70°F within 2 hours in the ice-water bath and then continuing to cool it to 40°F in the refrigerator within the following 4 hours.

MASTITIS AND THE CMT

Anyone who breeds goats, whether or not they milk them, should have a California Mastitis Test (CMT) kit on hand. Even nondairy does can get mastitis, especially at weaning time, when their udders are engorged with milk. The CMT is inexpensive, easy to use, and accurate. Each kit includes a plastic paddle with four compartments, a bottle of CMT solution to mix according to package instructions, and a card that shows you how to interpret the test results.

Run a test just before milking or, in the case of post-weaning does, any time you suspect mastitis. Squirt 1 teaspoon (2 ccs) of milk into each of two compartments on the paddle. Add an equal amount of reconstituted CMT solution to each compartment and then swirl the paddle in a circular motion to mix the compartments' contents. Ten seconds after adding the solution, read the results. If there is little or no thickening, the test is negative. Despite what the test instructions say, very slight thickening is okay with goats (the test was designed for cows; that's why the paddle has four compartments). Moderate thickening is a warning sign, and, if the mixture gels, your goat has mastitis.

Method 2:

- Prepare an ice-water bath for cooling your milk.
- Place the raw milk in a double boiler.
- Heat the milk quickly until it reaches 165°F and then remove it from your heat source. Or heat milk quickly to 161°F, maintain that temperature for 15 seconds, and remove it from the heat source.
- Cool your milk as in Method 1.

Freezing Milk

When your goat produces more milk than your family or bottle babies can use, freeze some for later. Raw milk freezes better than pasteurized milk, but all milk can be safely frozen with minimal loss in quality if the milk is handled properly.

For best results, plan to store frozen milk in a non-frost-free freezer cranked up to its coldest setting. If your family's main freezer is a frost-free model, consider buying a second freezer for milk storage. Under long-term storage, a frost-free freezer's constant freeze-thaw cycle damages frozen milk, making it likely to separate when it's thawed.

Some people like to add a pinch of baking soda to each jar of milk and gently shake it up before cooling. They say that this helps keep frozen milk from separating when you thaw it out.

Let's assume you've cooled your milk in quart canning jars. Pour the contents of each jar into a sturdy zip-top quart-sized freezer bag and then set the bags in a towel-lined container in the freezer overnight. The next day, fill gallon-sized zip-top bags with two quart-sized bags of frozen milk each for long-term storage. Write the date on all of the bags with a permanent marker; that way, you know to use older bags first.

Raw milk frozen in a non-frost-free freezer set at its coldest setting should stay good for at least 6 months, often longer. If you use a chest freezer, periodically rearrange containers as you fill it to keep the oldest bags near the top, where you can use them first.

Thaw frozen milk at room temperature. Frozen milk is still good if it separates as it thaws. Run separated milk through a blender and use it immediately or shake it up and feed it to bottle kids, pets, chickens, or pigs.

Strange Milk

Most fresh goat's milk is a pure, white, tasty delight, but there are exceptions.

Pink Milk

Pink milk is a sign of mastitis, so if you notice strings of blood or blood clots in your strainer, run a home mastitis test or take a sample to your vet. But don't panic. Newly freshened does, especially first fresheners, often give pink milk for a week or two as their udders adapt to milking. A hard blow to the udder can cause bleeding, too.

Slightly bloody milk that isn't caused by mastitis looks yucky, but it's safe to feed to pets or bottle kids. Or you can use it yourself if you want to. Chocolate or fruity syrups cover its slightly coppery tang quite well.

Strain pink milk twice, through two different filters, and let it set in the refrigerator overnight. Most of the blood will sink to the bottom. When it does, pour everything but the sediment into another container and use the milk however you like.

Chunky Milk

Anytime you find chunks in your milk filter, run a CMT test. If the test is negative, don't worry, because small white chunks are usually calcium particles or clotted fat or, occasionally, tiny pieces of shed tissue. If you filter them out, the milk will be fine. However, discard the milk if it tests positive for mastitis.

Flossie is a Pygmy/Nigerian Dwarf-cross doe.

Tainted Milk

The feed and forage your goat eats and the scents she inhales can flavor her milk. One way to avoid off-flavors in goat's milk is to prevent does from rubbing themselves on stinky bucks in rut and keep them out of smelly barns. But according to the University of California Cooperative Extension's bulletin,

Lucy, a Nigerian Dwarf doe.

"Milk Quality and Flavor," 80 percent of the off-flavors in goat milk are feed-related, so you also need to eliminate plants known to flavor milk from your goat's diet, or at least feed them at least 5 hours before milking. Plants to avoid include rutabagas, mangels, sugar beets (and beet pulp), turnips, cabbage, horseradish, marigolds, kale, flax, elderberry flowers, mint, wood sorrel, chamomile, fennel, daisies, cress, wild garlic and onions, ragweed, mustards, yarrow, wild lettuce, buttercups, sneezeweed, pepperwort, cocklebur, buckthorn, bitterweed, and wild carrot.

Refrigerated milk stored in plastic containers or in open-top containers of any kind tends to pick up refrigerator flavors, too. Milk also contains an enzyme called lipase that, when overproduced, gives milk the goaty taste of goat cheese. Properly cooling milk keeps this enzyme in check, as does pasteurization.

A goat who has subclinical mastitis usually gives off-flavored milk, as do some does in heat or immediately after giving birth. And some does consistently give odd-tasting milk. Try not to buy a doe like this.

Carrots for Sweeter Milk

Some say that feeding carrots to dairy goats makes their milk taste sweeter. In addition, carrots have antioxidant qualities and are fine sources of thiamin, niacin, vitamin B6, folate, manganese, vitamin A, vitamin C, vitamin K, and potassium. Slice a carrot into small chunks to prevent choking and drop it in your goat's feed cup while you're milking. We've tried this. Our goats love it, and it works!

GOAT KEEPER'S NOTEBOOK

Milking Kit to Go

Pack your milking supplies in a plastic food-service bucket. Between milkings, keep the bucket in the house with a pillowcase pulled over the top to keep everything clean.

Inexpensive Milking Bucket

Instead of an expensive stainless-steel milking bucket, milk into a stainless-steel bowl from a yard sale or thrift shop. Or remove the handles from a stainless-steel saucepan. Both options work just as well.

Give the Goat a Sip

In some places, goats are traditionally given a few ounces of their own milk at each milking because milkers believe that this prevents mastitis. If you try it, instead of using a strip cup, use a dark-colored bowl and don't discard that milk. When you're finished milking, add a few more squirts and offer the milk to your doe. Does it work? We don't know, but we haven't had a case of mastitis in the 3 years we've done it, and our does love it.

Fan Away Flies

If you don't like using fly spray on your goats, use a large box fan to blow away flies while you're milking.

The Udder Cream

Try human hand creams from the dollar store instead of expensive dairy creams. They work just as well.

Right, Left, or from the Back

Most goat books suggest milking from the doe's right side, but either side works equally well. You can also milk from the rear. This is the standard position in many parts of the world, and it usually works well, too. The exception would be for does that poop while being milked. Fortunately, not many do.

GLOSSARY

agouti: a grizzled coloration resulting from light and dark banded hairs; common to Pygmy goats and breeds descended from them

achondroplastic: referring to a type of dwarfism characterized by a normal-sized torso, head, and neck, with short limbs; Pygmy Goats are achondroplastic dwarfs

anthelmintic: a product used to expel internal parasites, commonly referred to as a "wormer" or "dewormer"

band: (noun) a strong rubber band used to castrate kids; (verb) the act of using an elastrator to apply one of these bands

billy: an outdated term denoting an uncastrated male goat; today's goat fanciers and breeders strongly discourage the use of this term

blubbering: the vocalizations of a buck or buckling during sexual display; it often sounds like "*what-what-what*"

bolus: a large pill; also a gob of partially digested food that is regurgitated to be rechewed

browse: (noun) edible woody plants, such as twigs, saplings, and wild berry canes; (verb) the act of eating browse

buck: an uncastrated male goat

buckling: an uncastrated male kid

CAE: caprine arthritis encephalitis, a nontreatable disease caused by a retrovirus

caprine: having to do with goats

cobby: wide and strong; of a stocky build

concentrates: the nonforage portion of a goat's diet: grains, meals, and commercial goat feed

conformation: an animal's physical characteristics

crossbred: having parents of two different breeds

cud: a glob of regurgitated food that's rechewed and swallowed again

dam: an animal's female parent

dehorning: the grisly removal of horns from an older kid or an adult goat

dental palate (or dental pad): a goat's toothless upper palate

disbud: to destroy a very young kid's horn buttons by burning them with a hot iron

disbudding iron: the electric or fire-heated tool used to disbud young kids

dished: concave

doe: a female goat

doeling: a female kid

dorsal stripe: a stripe of darker color running along an animal's back, parallel with its spine

dosing syringe: a tool used to give an animal liquid medicine or wormer

drench: (noun) liquid medicine given orally; (verb) to administer a drench

dry doe: a doe between lactations that is not producing milk

dry off: to cease milking a doe

dystocia: birthing problems; difficulty giving birth

easy keeper: an animal that doesn't require a lot of feed to stay in good condition

Elastrator: a tool used to apply a heavy rubber band to a buckling's scrotum for castration

elf ear: a type of LaMancha goat ear; Mini LaManchas' elf ears are triangle-shaped and up to about an inch long

F1: the first generation of an outcross; for example, an F1-generation registered Mini LaMancha would have a Nigerian Dwarf parent and a full-size LaMancha parent or a Mini LaMancha parent and a full-size LaMancha parent

F2 (and so on): subsequent generations of registered ancestors; for example, an F4-generation registered Mini LaMancha would have four generations of nothing but registered Mini LaMancha ancestors

feral: free-roaming, seemingly wild animals descended from domestic stock rather than from true wild animals that have never been domesticated; feral goats exist in many parts of the world, particularly Britain, Australia, and New Zealand

first freshener: a doe that has kidded for the first time

fitting stand: a platform with a headpiece to secure goats for grooming or other procedures

forage: fibrous animal feeds, such as browse, grass, and hay

freshen: to give birth and come into milk

gopher ear: a type of cartilage-free LaMancha ear with a tiny bit of skin around the ear opening

guard dog (or llama, or donkey): an animal that bonds with and stays with goats to guard them from predators, such as coyotes, dogs, wolves, bears, cougars, and eagles

halter: headgear used to lead or tie an animal

heat: the period when a doe is receptive to a buck and can conceive

herd book: a list of registered animals maintained by a registry

horn buds: two tiny lumps from which kids' horns emerge

intramuscular (IM) injection: an injection inserted into muscle

intravenous (IV) injection: an injection inserted into a vein

kid: a baby goat of either sex

kidding: giving birth to kids

lactation: the period during which a doe produces milk

lead: (noun) a rope or leash used to lead or tie an animal

livestock guardian dog (LGD): a dog of livestock-guardian breed heritage that lives with goats and protects them from predation

mL (milliliter): a unit of fluid medication measure equal to one cc

milking stand: an elevated platform fitted with a head stanchion; a doe stands up on the platform to be milked

milking through: milking a doe for more than 1 year without rebreeding her

nanny: an outdated word denoting a female goat; today's goat fanciers and breeders strongly discourage the use of this term

open doe: a doe that isn't pregnant

paddock: a small pasture

papered: registered

papers: a term referring to an animal's registration certificate

polled: naturally hornless

precocious milker: a doe that produces milk without first being bred

purebred: an animal whose ancestors for a set number of generations were all registered members of the same breed

registered goat: a purebred goat whose pedigree and particulars are registered in a registry's official herd book

Roman-nosed: possessing an arched or convex face

ruminant: a cud-chewing animal with a four-compartment stomach

ruminate: the act of chewing, particularly of chewing cud

rut: the period of time during the fall and early winter when bucks engage in courting behaviors such as urinating on themselves, bellowing, and blubbering

scur: a misshapen, unattached horn caused by improper or failed disbudding

sire: an animal's male parent

stanchion: a head restraint used to secure does while milking them

subcutaneous (SQ) injection: an injection inserted directly under the skin

udder: the female mammary system

urinary calculi: stones formed in the urinary tract

WAD: an acronym standing for West African Dwarf goats

wattles: a pair of fleshy, tubular appendages dangling from some goats' throats

wether: (noun) a castrated male goat of any age; (verb) the act of castrating a male goat

yearling: a goat of either sex between 1 and 2 years of age

RESOURCES

Books

Belanger, Jerry. *Storey's Guide to Raising Dairy Goats: Breeds, Care, Dairying*. North Adams, MA: Storey Publishing, 2000.

Boldrick, Lorrie. *Pygmy Goats: Management and Veterinary Care*. Orange, CA: All Publishing Company, 1996.

Carroll, Rikki. *Home Cheese Making: Recipes for 75 Delicious Cheeses*. North Adams, MA: Storey Publishing, 2002.

Damerow, Gail. *Fences for Pasture and Garden*. North Adams, MA: Storey Publishing, 2011.

Damerow, Gail. *Your Goats: A Kid's Guide to Raising and Showing*. North Adams, MA: Storey Publishing, 1993.

Grant, Jennie P. *City Goats: The Goat Justice League's Guide to Backyard Goat Keeping*. Seattle, WA: Skipstone, 2012.

Hall, Alice. *The Pygmy Goat in America with the Nigerian Dwarf*. San Bernardino, CA: Hall Press, 1982.

Holmes, Stephen. *The Book of the Goat*, 4th ed. London: F. Phillips, 1910.

Kubik, Rick. *Farm Fences and Gates: Build and Repair Fences to Keep Livestock In and Pests Out*. Minneapolis, MN: Voyageur Press, 2014.

Smith, Cheryl K. *Goat Health Care: The Best of Ruminations 2001–2007*. Cheshire, OR: Karmadillo Press, 2009.

Smith, Cheryl K. *Raising Goats for Dummies*. Hoboken, NJ: Wiley, 2010

Stewart, Patricia Garland. *Personal Milkers: A Primer to Nigerian Dwarf Goats*. Ashburnham, MA: Garland Stewart Publishing, 2008.

Weaver, Sue. *The Backyard Goat: An Introductory Guide to Keeping and Enjoying Pet Goats, from Feeding and Housing to Making Your Own Cheese*. North Adams, MA: Storey Publishing, 2011.

Weaver, Sue. *Hobby Farms Goats: Small-Scale Herding for Pleasure and Profit*. Irvine, CA: i-5 Publishing, 2011.

Weaver, Sue. *Storey's Guide to Raising Miniature Livestock*. North Adams, MA: Storey Publishing, 2009.

Weaver, Sue, Ann Larkin Hansen, Cheri Langlois, Arie McFarlen, Chris McLaughlin. *Hobby Farm Animals: A Comprehensive Guide to Raising Chickens, Ducks, Rabbits, Goats, Pigs, Sheep, and Cattle*. Irvine, CA: i-5 Publishing, 2015.

Breed Organizations

American Dairy Goat Association
www.adga.org
828-286-3801
Registers full-size dairy goats and Nigerian Dwarfs

American Fainting Goat Organization
www.americanfaintinggoat.com
936-201-0907

American Goat Society
www.americangoatsociety.com
830-535-4247
Registers Nigerian Dwarfs

American Nigerian Dwarf Dairy Goat Association
www.andda.org
434-738-8527

American Nigora Goat Breeders Association
http://nigoragoats.homestead.com

Australian All Breeds Miniature Goat Society
www.
australianallbreedsofminiaturegoatsociety.com
(07) 5462 7738

International Dairy Goat Registry–International Fiber Breed Registry
www.idgr-ifbr.com
202-570-IDGR

International Fainting Goat Association
www.faintinggoat.com
724-843-2084

Kinder Goat Breeders Association
www.kindergoatbreeders.com

Miniature Dairy Goat Association
www.miniaturedairygoats.net
360-225-1938

Miniature Goat Breeders Association of Australia
www.miniaturegoatbreedersassociation.com.au
(07) 5547 7217

Miniature Goat Registry
www.tmgronline.com
619-669-9978

Miniature Silky Fainting Goat Association
www.msfgaregistry.com
540-423-9193

Myotonic Goat Registry
www.myotonicgoatregistry.net
205-425-5954

National Miniature Goat Association
www.nmga.net

National Pygmy Goat Association
www.npga-pygmy.com
425-334-6506

Nigerian Dairy Goat Association
www.ndga.org
260-307-1984

Pygmy Goat Club (UK)
www.pygmygoatclub.org
01782 788225

Pygora Breeders Association
www.pygoragoats.org
315-678-2812

Equipment Suppliers

Caprine Supply
www.caprinesupply.com
800-646-7736
Goat gear, books, and milking and cheesemaking supplies; free print catalog

Hamby Dairy Supply
http://hambydairysupply.com
800-306-8937

Dairying Equipment

Hoegger Supply Company
http://hoeggerfarmyard.com
800-221-4628
Goat gear, books, and milking and cheesemaking supplies; free print catalog

Horseware Ireland USA
http://shop.horseware.com
800-887-6688

Horsewear Goat Coat
Jeffers Livestock Supply
www.jefferslivestock.com
800-533-3377
Vaccines, wormers, and farm supplies; free print catalog

New England Cheesemaking Supply Company
www.cheesemaking.com
413-397-2012
Cheesemaking supplies; free e-book

Pet Edge
www.petedge.com
800-738-3343
Discount prices; free print catalog

Port-a-Hut
www.port-a-hut.com
800-882-4884
Quonset-style portable housing for goats

Premier1 Equipment
www.premier1supplies.com
800-282-6631
Goat/sheep supplies and fencing supplies; free print catalogs

Valley Vet Supply
www.valleyvet.com
800-419-9524
Vaccines, wormers, and farm supplies; free print catalog

Feed Suppliers

Chaffhaye
http://chaffhaye.com
915-964-2406

Bagged fermented alfalfa hay
Purina
www.purinamills.com
800-227-8941
Purina hydration hay (listed under horse products)

US Alfalfa
www.usalfalfa.net
620-285-7777
Bagged alfalfa hay

General Information

Amber Waves Pygmy Goats
www.amberwavespygmygoats.com/library
Click on "Library" for hundreds of articles about goat keeping

American Consortium for Small Ruminant Parasite Control
www.wormx.info
Essential information about worms and worming

Dairy Goat Journal Library
www.dairygoatjournal.com/library
Click on "Library" for a collection of archived articles about goat keeping

FAMACHA video
www.youtube.com/watch?v=15TGB3CmIJc
The FAMACHA method of checking goats' eyes for anemia

Fias Co Farm
http://fiascofarm.com
Comprehensive information about goats and home dairying

Goat Blanket Instructions by Maxine Kinne
www.dairygoatjournal.com/85-1/maxine_kinne
Plans for sewing mini-goat coats

The Goat Mentor
www.greengablesmininubians.com/thegoatmentor.html
www.youtube.com/user/TheGoatMentor
Articles, videos, and services for goat keepers at all levels of experience

Jack and Anita Mauldin
www.jackmauldin.com/symptoms.html
A helpful chart of symptoms of goat diseases

Kinne's Minis
http://kinne.net
A huge collection of goat articles and links

Langston University Research and Extension Goat Dewormer Chart
www2.luresext.edu/goats/training/GoatDewormerChart.pdf
A dosing chart calibrated for minis as well as full-size goats

Maryland Small Ruminant Pages
www.sheepandgoat.com
The Internet's most comprehensive collection of goat links

Milking Stand Plans (Dairy Goat Journal)
www.dairygoatjournal.com/85-3/melissa_thomas
Mini-goat milking stand plans

Milking Stand Plans (Fias Co Farm)
http://fiascofarm.com/files/Milk_Stand_Plans.pdf
Adjusts to fit miniature or full-size goats

Pygmy Goat Weight Measuring Chart (Kinne's Minis)
http://kinne.net/weights.htm
Instructions and charts to help you accurately tape-weigh your goats

Tube Feeding Neonatal Small Ruminants
http://cru.cahe.wsu.edu/CEPublications/eb1998/eb1998.pdf
If you raise kids, you need this free publication from Washington State University

USDA Cooperative Extension Service
www.csrees.usda.gov/extension
Find your nearest extension office

Miscellaneous Websites

Craigslist
www.craigslist.org/about/sites

eBay
www.ebay.com

etsy
www.etsy.com

Facebook
www.facebook.com

Freecycle
www.freecycle.org

Pasteurization Information

Home Pasteurization of Raw Milk
http://pubstorage.sdstate.edu/AgBio_Publications/articles/ExEx14054.pdf

Raw Goat milk vs. Pasteurized Cow Milk
www.dairygoatjournal.com/83-4/maurissa_einsiedel

Periodicals

Dairy Goat Journal (print)
www.dairygoatjournal.com
800-551-5691

Dwarf and Mini Magazine (electronic)
www.facebook.com/dwarfandmini
United Caprine News (print)
www.unitedcaprinenews.com
817-297-341

Poisonous Plant Resources

Cornell University Department of Animal Science: Toxic Plants and the Common Caprine
www.ansci.cornell.edu/plants/goatlist.html

Fias Co Farm: Edible and Poisonous Plants to Goats
http://fiascofarm.com/goats/poisonousplants.htm

National Pygmy Goat Association: Poisonous Plants and Toxic Substances
www.npga-pygmy.com/resources/health/poisonous_plants.asp

INDEX

Note: Page numbers in **bold** typeface indicate a photograph.

C

D

E

F

G

H

I

J

K

L

M

N

O

P

Q

R

S

T

U

V

W

Z

PHOTO CREDITS

Front cover: Volt Collection/Shutterstock; Back cover: Anneka/Shutterstock

africa924/Shutterstock, 72
alison1414/Shutterstock, 112
Dave Allen Photography/Shutterstock, 92
Anneka/Shutterstock, 134
Jason Beals, 30
Bildagentur Zoonar GmbH/Shutterstock, 95
Kevin Castropassi/Flickr, 1, 130
Cliff/Flickr, 160
ClubhouseArts/Shutterstock, 61
Creamcup Minis, 152 (top)
Andrea Davis, 24 (bottom, inset)
Kim Depp, 16, 29 (both)
Jody Dillon, 155
Eliya Elmquist, 22 (bottom), 114, 152 (bottom)
Steven Frame/Shutterstock, 58
Dmitri Gomon/Shutterstock, 110
Jennie Grant, 77 (both)
Horsewear Ireland, 78
Malachi Jacobs/Shutterstock, 71, 89 (top)
JGade/Shutterstock, 138
Wolfgang Kruck/Shutterstock, 91
Sarah Marchant/Shutterstock, 161
Pete Markham/Flickr, 125
Scandphoto/Shutterstock, 64
Jen Schurman 27 (top)
Shmizla/Shutterstock, 86
Cheryl K. Smith, 24 (top), 150
Taviphoto/Shutterstock, 144 (top)
Robbie Taylor/Shutterstock, 122
Shannon Torgerson, 27 (bottom)
John and Sue Weaver, 4, 6, 8–14, 17, 18–20, 22 (top), 23, 24 (bottom), 26, 31, 32, 34–37, 39–42, 44, 46–47 (all), 49–54, 56, 60, 63, 65–69, 75, 76, 79–81 (both), 82, 84, 88, 89 (bottom), 90, 94, 96, 98, 100, 102, 104–109, 111, 116, 118 (all), 120, 123, 124, 126–129 (both), 131, 132, 136, 137 (all), 139, 140, 142–144 (bottom), 146–148, 154, 156, 157, 159
Peter Zijlstra/Shutterstock, 70
Cari Zisk, 21, 25

ABOUT THE AUTHOR

Sue Weaver is the author of twelve books, including five books in the Hobby Farms series (*Goats, Sheep, Llamas and Alpacas, Chickens,* and *Mini Goats*), and innumerable articles about livestock and poultry as well as a co-author of *Hobby Farm Animals: A Comprehensive Guide to Raising Beef Cattle, Chickens, Ducks, Goats, Pigs, Rabbits, and Sheep.* She and her husband are avid photographers. They live on a small ridge-top farm in the southern Ozarks, where they keep a wide assortment of animals, including Swedish Flower hens, several breeds of goat, Classic Miniature Cheviot sheep, horses, a guardian llama, and a pet razorback hog.